Edward Hazen

Popular Technology

Professions and Trades

Edward Hazen

Popular Technology
Professions and Trades

ISBN/EAN: 9783743686793

Printed in Europe, USA, Canada, Australia, Japan

Cover: Foto ©berggeist007 / pixelio.de

More available books at **www.hansebooks.com**

POPULAR TECHNOLOGY;

OR,

PROFESSIONS AND TRADES.

BY EDWARD HAZEN, A.M.,

AUTHOR OF

"THE SYMBOLICAL SPELLING-BOOK," "THE SPELLER AND
DEFINER," AND "A PRACTICAL GRAMMAR."

EMBELLISHED WITH EIGHTY-ONE ENGRAVINGS.

IN TWO VOLUMES.
VOL. II.

NEW YORK:

HARPER & BROTHERS, PUBLISHERS.

1870.

CONTENTS

OF

THE SECOND VOLUME.

	Page
The Musician, and the Musical Instrument Maker	7
The Sculptor	18
The Painter	29
The Engraver	42
The Copperplate Printer	51
The Lithographer	54
The Author	58
The Printer	63
The Type-Founder	73
The Stereotyper	77
The Paper-Maker, and the Bookbinder	81
The Bookseller	92
The Architect	97
The Carpenter	111
The Stone-Mason, the Brick-maker, &c.	114
The Painter, and the Glazier	129
The Turner	136
The Cabinet-Maker, and the Upholsterer	140
The Chair-Maker	149
The Carver, and the Gilder	153
The Cooper	157
The Wheelwright	161
The Potter	169
The Glass-Blower	178
The Optician	187
The Goldbeater, and the Jeweller	198
The Silversmith, and the Watchmaker	213
The Coppersmith, the Button-Maker, &c.	224
The Tin-Plate Worker, &c.	233
The Iron-Founder	242
The Blacksmith, and the Nailer	255
The Cutler	261
The Gunsmith	266
The Veterinary Surgeon	271

THE MUSICIAN, AND THE MUSICAL INSTRUMENT MAKER.

THE MUSICIAN.

1. THE word *Music*, in its modern application, has reference to the science which treats of the combination of sounds. It is founded upon the law of our nature, that every leading passion has its peculiar tone or note of expression understood by all human beings. Music, therefore, may be supposed to have been practised in the earliest ages; although it must have been a long time before it arose to the importance of a science.

2. According to the Mosaic records, Jubal, one of the descendants of Cain, played upon musical instruments, many hundred years before the flood. In the early period of the nations of antiquity, and in fact

among all semi-barbarous people of later periods, the character of poet and singer were united in the same individual; and the voice was frequently accompanied by musical instruments. The oldest song which has descended to our times, and which is stated to have been exhibited in this manner, was that sung by Miriam, the sister of Moses, on the occasion of the passage of the Red Sea by the children of Israel.

3. The Hebrews employed music in their celebration of religious worship, which consisted, in part, in chanting solemn psalms with instrumental accompaniments. It was also used by them on the occasion of entertainments, as well as in the family circle. It reached its greatest perfection amongst the Jews, in the days of David and Solomon. It is supposed, that the priests of Egypt were versed in music, before the settlement of the family of Jacob in that country; but how far the Israelites were indebted to them for a knowledge of this pleasing art, is altogether uncertain.

4. Music was held in very high estimation among the Greeks, who attributed to it incredible effects. They even assure us that it is the chief amusement of the gods, and the principal employment of the blessed in heaven. Many of their laws, and the information relative to the gods and heroes, as well as exhortations to virtue, were written in verse, and sung publicly in chorus to the sound of instruments.

5. It was the opinion of the philosophers of Greece, that music was necessary to mould the character of a nation to virtue; and Plato asserts, that the music of his countrymen could not be altered, without affecting the constitution of the state itself. But in his time and afterwards, complaints were made of the degeneracy in this art, and a deterioration of national manners through its influence. The degeneracy probably consisted in its application to the expression

of the tender passions; it having been previously applied, in most cases, to awaken patriotic and religious feeling.

6. The invention of music and of musical instruments, as in the cases of most of the arts and sciences among the Greeks, was attributed by the poets to some of the gods, or else to individuals of their own nation. It appears, however, from their traditions, that they received this art, or at least great improvements in its execution, from Phœnicia or Asia Minor. It began to be cultivated scientifically in Greece about 600 years before the advent of Christ.

7. The Romans seem to have derived the music which they employed in religious services from the Etruscans, but that used in war and on the stage from the Greeks. At an early period of their history, it was a great impediment to the progress of the art, that it was practised only by slaves.

8. The Roman orators pitched their voice, and regulated the different intonations through their speech, by the sound of instruments; and on the stage, the song, as well as part of the play itself, was accompanied with flutes. Wind-instruments of various kinds, comprised under the general name of *tibiæ*, and sometimes the cythera and harp, accompanied the chorus. In all these applications of music, the Romans had been preceded by the Greeks.

9. The Hebrews employed accents to express musical tones, but most other nations of antiquity used letters of the alphabet for this purpose; and, as they had not yet conceived the idea of the octave or parallel lines, to express a variety of tones in a similar manner by the aid of a key, they required a number of notes that must have been exceedingly perplexing.

10. The Greeks are said to have had about one thousand notes, half of which were for vocal, and the other half, for instrumental music. All these were

expressed by placing the letters of their alphabet, or parts of them, in different positions. Accents were also used, partly by themselves, and in connexion with the letters.

11. The lines of a poem, set to music, were placed under the letters expressing the tones. The letters for the instrumental part were placed first, and under them those for the voice. The notes of the Greeks and Romans were not required to indicate the time in which they were to be pronounced, since in general the syllables of their language had a natural and distinct quantity. In the cases in which there was a liability to mistake, the syllables were marked with A, if long, and with B, if short.

12. The Romans expressed the fifteen chief tones of the Greeks with the fifteen first letters of the Latin alphabet; and these were reduced to seven, by Pope Gregory I., towards the end of the sixth century; so that the first seven capital letters were used for the first octave, the small letters for the higher octave, and the small letters doubled, for the highest octave. Parallel lines were soon after invented, on which the letters were written.

13. Musical sounds were expressed in this manner until the year 1024, when, according to some authors, Guido Aretine, a monk of Arezzo, invented points and rhombuses. He also introduced the use of five parallel lines, upon and between which his notes were written. The seven letters which had formerly been used as notes, now became clifs.

14. Still, however, the means of determining the duration of sound belonging to each note, without consulting the quantity of syllables in the verses to be sung, were yet to be provided. This desideratum was supplied by one Franco, a German of Cologne, who lived towards the end of the eleventh century. Some, however, attribute this improvement to John de Murs

The division of one note into others of less value was invented, in the sixteenth century, by Jean Mouton, chapel-master to King Francis I. of France.

15. The knowledge of music, as a science, was preserved in Europe, after the overthrow of the Western empire, through the influence of the Church. The apostles, and Hebrew converts generally, had been accustomed to the sacred music of the Jews; and, on this account, it was easy to continue the use of the same psalms and hymns in the Christian Church.

16. Many of the Grecian and Roman melodies were also set to words adapted to Christian worship. In regard to the manner of singing, in the early days of the Church, it was sometimes in *solo*, sometimes in *alternate strains*, and at other times in *chorus;* in which the whole assembly joined, repeating what had been before sung or read. In the fourth century, with the view of securing the proper execution of this part of divine worship, *precentors* were instituted, who were considered regular officers of the Church.

17. Pope Gregory I., surnamed the Great, distinguished himself by establishing a new singing-school, which became a model for many others, in the western division of the Church. In consequence of these schools, the singing became more artificial; and this, together with the circumstance that the hymns were in Latin, which had become obsolete, at length excluded the people from any participation in this part of the public worship.

18. Gregory also made a selection of the existing songs of the Church, and introduced a *chant,* which, through his influence, and that of his successors, was at length extended throughout Europe. It received the appellation of the *Gregorian chant* from his name. It was also called the *choral song,* because it was sung by a choir. This chant is said to be the foundation of our present church-music.

19. Music, in distinct parts, was not known until after the introduction of the improved method of writing music, invented, as before stated, by Guido Aretine and Franco. The development of harmony, in four parts, was assisted by the *choral;* but it was more particularly advanced by musical instruments, and especially by the organ. In the fifteenth century, music began again to be treated scientifically.

20. The Reformation produced great changes in the character of sacred music. Before that event took place, this part of religious worship was confined to a few fixed forms of texts, as in the mass, and this is still the case in the Roman Catholic Church; but the Protestants allow great variety both in the poetry and music. Luther's agency in the production of these changes was very considerable. During the seventeenth and eighteenth centuries, church music became continually more brilliant, and always more corrupted, by the intermixture of profane music.

21. In the sixteenth and seventeenth centuries, there grew up, at the courts of the European monarchs, the free chamber style, from which arose that which was afterwards used in the theatre. The opera, which originated with three young noblemen at Florence in 1594, has contributed especially to the splendor and variety of modern vocal music, the advancement of which is claimed particularly by the Italians, as that of the instrumental kind is claimed by the Germans and French.

22. The composition of music, and its execution either vocally or instrumentally, as well as the business of imparting a knowledge of it to others, are embraced in the employment of the musician; although it is seldom, that all these branches are practised by one and the same individual. Music is one of the fine arts, and, during the middle ages, was one of the branches of what was then considered a learned education.

23. Since the scientific revival of music, the art has had so many distinguished professors, that we will not even attempt to give a list of their names. Their number was increased, and the art greatly perfected, by the singing-schools, called *conservatories*, established especially in Italy, either at the public expense, or by the liberality of individuals.

MUSICAL INSTRUMENT-MAKER.

1. This artist unites in his business some of the operations of the cabinet-maker, turner, and brazier. He also is dependent upon the wire-drawer, and the tanner and currier, for some of his materials. So great, however, is the number of musical instruments, and so different their nature and construction, that the business of making them is divided into several branches, all of which are never pursued, or carried on, by one person. But, without reference to the several divisions of this business, we will proceed to mention or describe the principal instruments which are now in most common use.

2. The *organ* is the largest of all musical instruments, and, in its improved state, so complex that a mere description of it cannot be well understood. Nevertheless, we will endeavor to give the reader some idea of the general principles on which it is constructed.

3. The most essential and prominent parts of this machine are the *wind-chest*, the *pipes*, and the *bellows*. The former of these is an oblong box, made perfectly air-tight, and placed in a horizontal position. The top of this chest is perforated with several rows of holes of different sizes, and into these are inserted the pipes. Those for the higher notes are of a cylindrical form, and are made of a mixture of metals, chiefly of tin and lead; but those designed for the expression of the lowest notes of the base are made

of wood, in a square form. The dimensions of these pipes are regulated by a *diapason*, or *scale*.

4. There are as many of these rows of pipes, which are called *stops*, as there are kinds of tones in the organ; and to every row or stop is a plug, attached to a slide, which is denominated a *register*, and which is designed to regulate the admission of wind into the pipes. The pipes are also furnished with valves, which can be opened at pleasure, by means of keys similar to those of the piano-forte. Some organs have few, others have many stops; and, in order to regulate the force of sound, most church organs have two or three rows of keys, whereby a greater or less number of pipes may be filled, and the powers of the instrument may be controlled in what is called the *small organ*, or let loose, so as to become the *full organ*.

5. The fingering of an organ is similar to that of the piano-forte, so far as relates to the position of the keys; but, on account of the great number of holding notes in organ music, and the manner in which the sound is produced, the fingers are more kept down; whence it is considered injurious for performers on the piano-forte to practise on the organ, lest that lightness of touch, so necessary for the former instrument, be affected. It is hardly necessary to remark that, during the performance on the organ, the wind-chest is filled by means of the bellows.

6. The structure of the organ is lofty, elegant, and majestic; and its solemnity, grandeur, and volume of tone, have obtained for it a pre-eminence over every other instrument for the sacred purposes to which it has been applied. The largest organ known is in St. Peter's Church, at Rome. It has one hundred stops.

7. The church organ was probably suggested by the *water organ* of the Greeks, which was invented

five or six hundred years before our era. At what period, organs began to be employed in churches, cannot now be ascertained. By some, it is said that Pope Vitelianus caused them to be used in Rome in the seventh century. Others are of opinion, that they were not introduced until three hundred years later. But, be this as it may, the church organ was not in common use until the fourteenth century; and now it is very different in its construction from that of early times. It has received many additions and improvements since the beginning of the fifteenth century.

8. The *hand* or *barrel organ* consists of a moveable cylinder, on which, by means of wires, pins, and staples, are marked the tunes which it is intended to perform. These pins and staples, by the revolution of the barrel, act upon the keys within, and give admission to the wind from the bellows to the pipes. The hand organ is so contrived that the revolution of the barrel gives motion to the bellows.

9. There are several instruments belonging to the class of *horns*, all of which are made of brass or silver. Those of the latter kind of metal are by far the softest in tone, but brass is the material most commonly employed. The chief instruments belonging to this class are the trumpet, the French horn, the bugle, the Kent bugle, the trombone, and the basshorn. The *serpent* seems to be the connecting link between the trumpet and the flute.

10. The instruments classed with the flute, are the common flutes of various keys, German flutes, and several kinds of flageolets. Nearly allied to these are the clarionet, the hautboy, and bassoon. The breath is applied to the flageolet through an ivory tube at the end; and, in the three last named instruments, a thin reed, capable of a free vibration, is a part of the mouth-piece.

11. Of the instruments which produce musical sounds by the vibration of strings, there are a great number, of which the following are the principal;—the lyre, the harp, the guitar, the lute, the dulcimer, the harpsichord, the spinnet, the piano-forte, the violin, the violincello, and the base-viol. The strings of the three last are agitated with a bow; but those of this class first mentioned, are vibrated by the thumb and fingers, by some little instrument held in the hand, or by little hammers, moved by keys, as in the piano-forte.

12. The *piano-forte* is said to be the invention of Gottlieb Schroder, of Hohenstein, in Saxony, born in Dresden, about the year 1717. Before the introduction of this instrument, the clavichord, harpsichord, and spinnet, supplied its place. On all of these instruments complete harmony can be produced by a single performer, and the most difficult series of tones can be executed with rapidity, by means of a simple mechanism.

13. The *piano-forte* has been gradually improved, until it has become one of the most elegant instruments in the whole compass of musical practice. In firmness and strength of tone, the English piano-fortes formerly surpassed all others; but, within a few years, they have been equalled, and in some respects excelled, by those of American workmanship. The manufacture of this instrument constitutes the most extensive branch of musical instrument-making.

14. The instruments of percussion are the military drum, base-drum, kettle-drum, tabor, tamborine, and the triangle. The kettle-drum has received its name from its conformation. It has but one head, and is used in orchestres, and by the cavalry of modern armies, especially in Europe. The tabor has two heads, about three inches apart, and is beaten with one stick. The tamborine has one head, drawn over a hoop, to

which are attached small bells and bits of tin, to make a jingling sound. The time is beaten on the head with the hand.

15. The *bag-pipe* is a wind instrument of high antiquity among the northern nations of Europe; but it has been so long a favorite with the natives of Scotland, that it may be considered their national instrument. It consists of a leather bag and three pipes. The first of the pipes is that by which the droning noise is produced, the second emits wind from the bottom of the bag, and the third is that on which the music is made.

16. During the performance on the bag-pipes, the bag is placed under the arm, and worked like a bellows, while the notes are modulated as on a flute or hautboy, by stopping and opening the holes, nine in number, with the ends of the fingers and thumb. The bag is filled by means of the breath blown into it through a pipe. In Rome, at the time of Advent, the peasants of the mountains express their veneration for the Virgin by playing on this instrument before her image.

THE SCULPTOR.

1. SCULPTURE is one of the fine arts. In its most extended sense, it includes not only modelling figures in clay, wax, and plaster of Paris, and carving them in wood, stone, and marble, but also *casting* them in bronze, lead, or iron, as well as enchasing and engraving.

2. The productions of this art are known under various denominations, but the principal are *statues*, *busts*, and *bas-reliefs*. The first of these are entire representations of men or animals in full relief; the second are upper parts of statues; and the last are figures more or less elevated from the body or ground on which they are formed.

3. The different degrees of elevation in reliefs, are expressed by various terms borrowed from t' e Ital

ian. A figure is said to be in *alto relievo,* or *high relief,* when but a small proportion of it is buried in the back-ground; in *mezzo relievo,* or *middle relief,* when one half of it is above the surface; and in *basso relievo,* or *low relief,* when but little elevated, like figures upon coin. Bas-reliefs are usually applied as ornaments to buildings, and to the pediments of statues.

4. The subjects of sculpture, with a few exceptions, are the same as those of painting; and the course of study essential to proficiency in either, is very similar. They both require much taste and practice, and a thorough knowledge of the human form and other objects frequently represented. The young artist begins with imitating the most perfect models of Grecian art; and, after having become well acquainted with their beauties, he proceeds to the imitation of nature.

5. When any considerable work in stone or marble is to be done, the sculptor forms a model of clay or wax, to guide him in the execution. The soft material is moulded to the proposed form with the hands and small instruments of ivory. The model is by far the most difficult part of the work, and it is here the genius of the artist is to be displayed. The process of copying the model in stone or in any other substance, is an operation merely mechanical, and can often be done by another person as well as by the scientific sculptor himself.

6. The model having been prepared, the block of marble or stone is marked at certain points corresponding to its chief elevation and concavities. The material is then wrought to the rough outline of the figure, by means of strong steel points, drills, and other perforating tools; and the asperities are afterwards removed with chisels, and with rasps and files of different shapes. When a high polish is required, it is produced by friction with pumice-stone, tripoli, and straw ashes.

7. Marble and stone are carved in a similar manner; but the latter, being softer, can be wrought with less difficulty. The defects which may be met with in the stone are repaired with a composition of plaster of Paris and the same stone, pulverized and mixed with water.

8. *Casts* in plaster of Paris and bronze are taken from models, statues, busts, bas-reliefs, and living persons. To do this, it is necessary to form a mould from the subject to be copied. This is done by spreading over it some soft substance, which can be readily forced into all the cavities, and which will harden by drying or cooling. Plaster of Paris is the most usual material employed for this purpose.

9. When the subject is a bas-relief, or any other one-sided figure of a similar kind, the mould can be withdrawn without injury, in a single piece; but if it is a statue, or any other figure of like form, it is necessary to divide the mould into several pieces, in order to a safe removal. These pieces again united constitute a perfect mould. While the artist is forming the mould on the face of a living person, the latter breathes through tubes inserted into the nostrils.

10. In taking casts from such a mould, the internal surface is oiled to prevent adhesion, and then plaster mixed with water is poured into it through a small orifice. The mould is afterwards turned in every direction, that the plaster may cover every part of the surface; and when a sufficiency of it has been distributed to produce the requisite strength, and the plaster has acquired the proper solidity, the several pieces are removed from the cast, which, of course, is an exact resemblance of the subject on which the mould was formed.

11. Superfluous portions of the material, produced by the seams in the mould, are removed with suitable instruments, and applications of fresh plaster are

made, where necessary to repair blemishes. The cast is finished by dipping it in a varnish made of soap, white wax, and water, and afterwards rubbing it with soft linen. The polish produced in this manner, approaches that of marble.

12. The durability of plaster casts, exposed to the weather, is greatly increased by saturating them with linseed oil combined with wax or rosin. They are made to resemble bronze by the application of a soap composed of linseed oil and soda, and colored with the sulphate of copper and iron.

13. Moulds are, also, formed of a warm solution of glue, which hardens upon cooling, and such are called *elastic* moulds. This material is sometimes preferred on account of its more easy separation from irregular surfaces. For small and delicate impressions in bas-relief, melted sulphur is sometimes employed; also a strong solution of isinglass in proof spirits. All three of the substances last mentioned yield sharper impressions than plaster of Paris.

14. Statues designed to occupy situations in which they may be exposed to the weather and mechanical violence, are often made of bronze cast in moulds. The external portions of the mould are made on the pattern, out of plaster, brick-dust, and water. The mould is then covered on the inside with a coating of clay as thick as the bronze is intended to be, and the several pieces are afterwards put together, or *closed.* The internal cavity is next filled with a composition like that on the other side of the clay.

15. When this has been done, the several pieces forming the outside of the mould are separated, and the clay carefully removed. These having been again united, and the core or internal portion of the mould secured in its true position, the whole is bound with iron hoops, and thoroughly dried. The melted bronze is poured into the cavity formed by the remo-

val of the clay, through an aperture made for the purpose. The cast is afterwards rendered smooth by mechanical means.

16. It is conjectured with much reason, that sculpture was one of the arts practised before the deluge, and that it was transmitted to posterity by the survivors of that catastrophe. The first images were probably made for the purpose of perpetuating the memory of the dead; but, in process of time, they became objects of adoration. As the Chaldeans were unquestionably the first idolators after the flood, so are they supposed to have been the first who made progress in sculpture.

17. The first notice of this art in the Mosaic writings, is found in the passage relative to the teraphim, or idols, which Rachel, the wife of Jacob, carried clandestinely from her father's house; and the first persons mentioned in the Bible, as artists, are Aholiab and Bezaleel, who formed the cherubim which covered the mercy-seat, together with some other furniture of the tabernacle, and the sculptured ornaments of the garments of the high-priest.

18. From the same authority, we learn that the nations expelled from Canaan, by the Jewish people, were not ignorant of sculpture and painting; for Moses repeatedly commands the latter to destroy the *pictures* and molten images which might be discovered in their progress through the land. The Israelites crossed the river Jordan about 1500 years before the commencement of our era.

19. From this time to the end of the Jewish polity, we often meet in the Scriptures with indications of the fine arts; but the splendor of Solomon's temple, clearly points out the days of that prince as the period in which they had attained their greatest perfection in Judea.

20. The Babylonians, Assyrians, and Phœnicians,

became considerably skilful in sculpture, at a very early period, as we learn from early history, and some existing remains. The same remark is also applicable to the inhabitants of Hindostan. But writers have been more particular in noticing the style of design among the Egyptians, because the progress of the arts among that people is more easily traced, and because it is supposed to elucidate that of most other ancient nations.

21. The chief objects of sculpture, among the Egyptians, were pillars, and other architectural ornaments, idols, the human figure, animals, and hieroglyphics, engraved in a kind of bas-relief on public edifices, and the forms of animals. Most of the great works of this nation are supposed to have been executed during and after the reign of Sesostris, who lived in the days of Rehoboam, king of Israel, or about 1000 years before the Christian era.

22. But of all the nations of antiquity, the Greeks were the most distinguished for sculpture. They derived the first rudiments of the art from the Phœnicians, or Egyptians, although they assert that they themselves were its inventors. Its existence, in a rude state, among that people, preceded that of letters or scientific architecture.

23. Dædalus, who lived about 100 years after Moses, was the first sculptor among the Greeks, of any notoriety. The statues made before his time, were stiff, formal figures, having the arms attached to the body, and the legs united, like the mummy-shaped productions of Egyptian art. He separated the legs of his statues, and placed them, and the upper extremities, in a natural position. He also was the first sculptor who made the eyes of his statues open. On account of these improvements, the Greeks said, that his divine genius made statues walk, and see, and speak.

24. The disciples and imitators of Dædalus were called his sons, and artists, generally, *Dædalides.* Soon after this period, schools of design were established in the island of Ægina, at Corinth, at Sicyon, and in Etruria, in Italy: but it seems that no good representations of the human form were effected until near the time of Phidias, who was born 444 years before Christ.

25. This most distinguished of all the votaries of sculpture, flourished at or near the same time with the dramatic poets, Æschylus, Euripides, and Sophocles; the philosophers, Socrates, Plato, and Anaxagoras; and the statesmen and commanders, Pericles, Miltiades, Themistocles, Cimon, and Xenophon. This was the most refined period of Grecian history, and of all others, the most favorable in its moral and political circumstances, for the development of genius.

26. Phidias was the author of the *ideal style*, which, in the fine arts, may be defined, the union of the perfections of any class of figures. Among the distinguished productions of this artist, the colossal statues of Minerva and Jupiter Olympius, made of gold and ivory, have excited the greatest astonishment. The former, executed for the Parthenon of Athens, was twenty-six cubits in height; and the latter, for a splendid temple at Elis, was about the same height, although seated upon a throne.

27. The favorite disciples of Phidias, were Alcamenes, of Attica, and Agoracritus, of Paros; and at the same time with them, flourished Polycletus, of Argos, Miron, of Bœotia, and Pythagoras, of Rhegium. The *beautiful style* soon succeeded to the ideal; the authors of which, were Praxiteles and Scopas, who brought the art to the highest perfection,—since, in their productions, they united beauty and grace. After the days of these two artists, sculpture began to decline; although it continued to be practised with

considerable success, for some centuries after this period.

28. The great superiority of the Greeks in the art of sculpture, is ascribed to various causes; among which are classed, their innate love of beauty, and their own elegance of form, combined with the frequent opportunities of studying the human figure, in places where youth were in the habit of performing athletic exercises in a state of nudity. To these may be added, the practice of awarding to citizens a statue of their own persons, for eminent services to the state, and for excelling in exercises at the public games.

29. The fine arts were nearly extinguished in Greece, by the conquest of the Romans; who, with ruthless rapacity, seized upon, and transferred to their metropolis and villas, the superb works of taste with which the country abounded. By these means, however, a taste for the arts was produced among the Romans, who encouraged with great liberality the Greek artists who resorted in great numbers to their city.

30. The arts at length declined at Rome, and finally became nearly extinct in that city, soon after Byzantium was made the capital of the Roman empire, in 329 of the Christian era. The new capitol was enriched by the most valuable statuary of the old metropolis, and by a farther pillage of Greece. Artists were also encouraged with a munificence similar to that of former times; and many new subjects in painting and sculpture, in illustration of the Christian scriptures, were executed as embellishments for the sacred buildings of the city.

31. The art of sculpture necessarily declined during the time of the unsettled state of Europe, which followed the conquests by the barbarous nations. It, however, was not altogether lost, but was occasion-

II.—C

ally practised, although in a very rude manner, in several kingdoms of Europe. In the eleventh century, after the terrors of the northern invasions had passed away, and the governments had become more established, the arts of design began a regular course of improvement, which has been denominated their revival.

32. This improvement was promoted by means of the frequent intercourse which had sprung up between the commercial cities of Italy and the Greek empire. In 1016, the Pisans founded their great church, called the Dome of Pisa; and, in its construction, they employed many noble pillars and other fragments of Grecian edifices. They also engaged upon the work several Grecian sculptors and painters, who exerted in their service the little skill which had come down from antiquity.

33. The specimens of ancient art thus introduced at Pisa, and the works of these artists, at length incited several Italians to emulation; among whom was Nicolo Pisano, who became the restorer of true taste in the arts, in the thirteenth century. At this period, the crusades had diffused such a zeal for the Christian religion, that magnificent churches were built in every part of Italy, in the designing of which, and in their decoration with sculpture, Pisano and his scholars were universally employed.

34. John Pisano, the son of Nicolo, was also an architect and sculptor of eminence; and by him was built, for King Charles, a castle, and several churches, at Naples. He also executed several pieces of sculpture, and superintended the construction of some edifices in Tuscany. This sculptor, who died in 1320, had several pupils, of whom Agostino and Agnolo Sanesi were the best sculptors of the time.

35. In 1350, an academy of design was formed at Florence by the union of several painters, sculptors,

and architects. This institution was called after St. Luke, whom tradition makes a painter by profession. The society was afterwards munificently patronised by the Medici, a noble and wealthy family of that city.

36. From this school, there soon proceeded a great number of skilful artists, among whom were the sculptors Lorenzo Ghiberti, Donatello, and Brunileschi; and after these, others perhaps still more distinguished, until it produced Michael Angelo Buonarotti, who, as a universal genius in the arts of design, has excelled every other artist, whether ancient or modern.

37. This great man was born in Florence, in 1474. His father, having discovered his talent for designing, made him a pupil of Dominic Ghirlandaio, who instructed him in the first principles of the art of draw ing. He studied statuary under Bartoldo; and, in his sixteenth year, copied the head of a satyr in marble, to the admiration of all connoisseurs. On account of his great promise, he was liberally patronised by Lorenzo de Medicis, who, besides allowing him a pension, gave him a lodging in the palace, and a place at his table. After the death of this prince, he enjoyed the same favors from his son, Pietro de Medicis.

38. His reputation as an artist having been established at Florence, he was called to Rome by Julius II. From this time, he remained chiefly in the service of the popes, for whom he executed many inimitable works, both of sculpture and painting. He was also an architect of the first order; and, as such, was employed on St. Peter's Church, as well as on several other public edifices. He died in 1564, at an advanced age.

39. Sculpture, having been brought to as high a state of perfection as it was ever likely to be carried, began to decline in Italy, as it had done before, under similar circumstances, in ancient times; but as barbarism did not again occur to overwhelm it, it did not

entirely disappear. It continued to be practised, although in a very inferior degree, until it was again revived by Antonio Canova, near the close of the eighteenth century.

40. The French nation, from its vicinity and intercourse with Italy, obtained from that country the means of improvement in every branch of the fine arts. Accordingly, native artists of considerable merit occasionally appeared. The kings of France, also, often employed Italian architects and sculptors on their great public works. In the reign of Francis I., Leonardo da Vinci, and two other artists from Italy, established a school of fine arts similar to that of St. Luke, at Florence; and the genius of the people, added to national munificence, have kept a respectable school of sculpture to the present time.

41. Considerable ability in sculpture has likewise been exhibited by native artists of Spain, Germany, Holland, England, and some other countries of Europe; but whatever skill has been displayed in any of these countries has been derived, in an indirect manner, at least, from Italy. In the United States, the fine arts have been cultivated with considerable spirit. An academy for this purpose has been established both in New-York and Philadelphia, and a picture gallery has been connected with the Athenæum in Boston, in which the annual exhibition of paintings is respectable.

THE PAINTER.

1. PAINTING is the art of representing visible objects, by means of lines and colors, on a plane surface, so as to produce the appearance of relief. It is justly ranked among the highest of that class of arts denominated fine, or liberal; and its tendencies and powers being similar to those of poetry, it is considered an employment worthy of men of the most exalted rank.

2. The theory and practice of this ingenious and delightful art, are divided by its professors into five distinct branches,—*invention, composition, design, chiaro-scuro,* and *coloring. Invention* relates to the choice of subjects to be introduced into a picture. It is this which gives the highest character to the artist, as it

affords the greatest opportunity to display the powers of his mind.

3. *Composition* regards the general distribution and grouping of figures, the choice of attitudes, the disposal of draperies, the situation of the scene itself, as well as the arrangement and connexion of the various parts of the scenery. Invention and composition are employed particularly in the first rough sketch of a picture.

4. *Design* refers to the expression of a proposed picture in simple contour, or outlines. It is applied in making the first rough sketch of the picture, whether in miniature or in its full size, as well as in the more accurate expression of the form of the figures, in its final finish. The artist, in making his design, is guided in drawing his lines by the rules of *perspective*, according to which he is able to *foreshorten* objects, and thereby diminish the space which they occupy, without giving them the appearance of diminished magnitude.

5. *Perspective* has been defined the art of delineating the outlines of objects on any given surface, as they would appear to the eye, if that surface were transparent, and the objects themselves were seen through it, from a fixed position. For example; when we look through a window at a mass of buildings, and observe that part of the glass to which each object, line, or point appears opposite, we find that their apparent position is very different from their real. A delineation of these objects on the glass, as they appear, would be termed a representation in perspective.

6. Correct perspective is the foundation of scientific painting; and, next in importance to this, is a proper distribution of light and shade. This branch of the art is called *chiaro obscuro*, or, when abridged, *chiaro-scuro*. The term is Italian in its origin, and

its literal meaning is *clear* and *obscure*. To the skilful management of light and shade, we are indebted for the strength and liveliness of pictures, and their relief, or the elevation which certain parts appear to assume above the plane upon which the objects are represented.

7. By the aid of perspective and chiaro-scuro, very good representations in one color are attained. Drawings in India-ink and crayons, as well as pictures taken from engraved plates and wood-cuts, are specimens of such productions. But a nearer approach to the appearance of nature, is made by the employment of colors analogous to those which are found to exist in the objects to be represented.

8. To produce various hues in painting, the artist employs coloring substances, which, either alone or by mixture, are analogous to them all; and, in their use, he is careful to apply them in such a manner, that the true colors remain distinct from the lights and shades necessary to produce the objects in relief. Artificial colors are divided into *warm* and *cold*. The former are those in which red and yellow predominate; the latter are blue, gray, and others allied to them.

9. Before coloring substances can be applied in painting, they must be reduced to extreme fineness, and be mixed with some tenacious fluid, to cause them to adhere to the surface on which they are to be spread. The fluid employed for this purpose, and the mode of applying the colors, have given rise to the different kinds of painting, of which the following are the principal: *crayon, water-color, distemper, fresco,* and *oil-painting*.

10. The most simple mode of applying the colors is by means of crayons. They are made of black lead, a species of chalk, or of a mixture of coloring matter with gum, size, or clay. For painting in *wa-*

ter-colors, the substances employed in communicating the tints are combined with gum, and formed into cakes or lozenges. When about to be used, they are dissolved in water, on glass or a glazed surface. The application in painting, is made by means of a camel's-hair pencil.

11. Painting in *distemper* is used for the execution of works on a large scale, such as stage scenery, and the walls of apartments. The coloring substances are mixed with water, rendered tenacious by size or solutions of glue, or by skimmed milk, increased in tenacity by a small quantity of thyme. Linseed or poppy oil often serves as a vehicle for the colors, in this kind of painting.

12. Paintings in *fresco* are executed on walls of plaster. The coloring matter mixed with water, being applied to the plaster while the latter is in a fresh state, sinks in, and incorporates itself with it, so as to become very durable. During the execution of the work, the plaster is applied to the wall in successive portions, no more being added at a time, than can be conveniently painted before it becomes dry. Works of this kind must be executed with great rapidity; and, on this account, patterns, called *cartoons*, are previously drawn on large paper, to guide the artist in his operations.

13. *Oil painting* derives its name from the mixture of the colors in oil. The oils used for this purpose are extracted from vegetables; and, on account of the rapidity with which they dry, are denominated drying oils. For most purposes, this mode of painting is decidedly superior to all others. It admits of a higher finish, as it allows the artist to retouch his works with greater precision. The colors also blend together more agreeably, and produce a more delicate effect. Oil paintings are executed on canvas, wood, or copper.

14. Paintings are imitated with surprising elegance, by cementing together colored pieces of glass and marble, as well as those of wood. Representations by these means, are called *Mosaics*, or *Mosaic paintings*. The cause of their having received this appellation cannot be ascertained. Some, without much reason, attribute the origin and name of this branch of the art to Moses. Others suppose that works of this kind have been thus denominated, because they were first employed in grottoes dedicated to the *Muses*.

15. Drawings and paintings are divided into classes, according to the nature of the objects represented, the principal of which are *historical, architectural, landscape, marine, portrait, still life, grotesque, botanical,* and *animal*. The subordinate divisions of these branches are very numerous.

16. The propensity to imitation, so deeply rooted in the human mind, is the foundation of the arts of design; and there can scarcely be indicated a lengthened period in the history of man, in which it was entirely inactive. It may have first been accidentally exhibited in tracing the form of some object in the sand; or resemblances in sticks and stones, may have originally suggested the idea of imitations by means of lines and colors.

17. Although painting and sculpture may be supposed to have existed, at least in a rude state, at a very early period, and even before the deluge, yet the reign of Semiramis, queen of Assyria, 2000 years before Christ, is the earliest to which authentic history extends. Diodorus Siculus relates that the queen, having thrown a bridge across the Euphrates, at Babylon, erected a castle at each end of it, and inclosed them with walls of considerable height, with towers upon them. The bricks of which they were constructed, were painted before they underwent the

fire, and were so put together, that single figures, and even groups of them, were represented in colors.

18. This author supposes also, that the arts had attained nearly an equal degree of cultivation about the same time in Egypt, sculpture, as best serving idolatrous purposes, being in both countries much in advance of the sister art of painting. But, in neither country, was painting or sculpture brought to a great degree of perfection.

19. In Egypt, independent selection of objects, and variety of exhibition, never appear to have been much regarded. When a specific form of character had been once adopted, so it remained, and was repeated unchanged for ages. Little action, and no expression, was given to figures. The chief employment of the Egyptian artists, seems to have been the painting of the chests of mummies, and the ornaments on barges and earthenware.

20. Painting, in the early days of its existence, was employed chiefly in the exhibition and preservation of historical facts; and, wherever it remained faithful to these objects, it was obliged to sacrifice the beautiful to the significant. Only in those countries where alphabetical writing existed, could painting elevate itself to a fine art.

21. The Pelasgi, who expelled or subdued the earlier inhabitants of Greece, and colonized that country, probably brought with them the rudiments of this art; and it at length grew up with its sister arts. In some of the stages of its progress, this intelligent people, no doubt, received useful hints from other countries, and especially from Egypt; yet they finally surpassed all the nations of antiquity in this branch of art.

22. The Greeks, with singular care, have preserved the names of their artists from the earliest periods of their practice. Ardens, of Corinth, Telephanes and Crato, of Sycion, and some others, are noticed as

such, when painting had advanced no farther than the mere circumscription of shadows by single lines.

23. The different kinds of painting, as marked by the successive stages of the art among the Greeks, are as follows; 1. The *skiagram*, or drawing in simple outlines, as in the circumscriptions of shadows. . The *monogram*, including both the outlines and others within them. 3. The *monochrom*, or picture in a single color. 4. The *polychrom*, or picture of many colors.

24. Although the names of the Grecian artists were carefully preserved, the time in which they lived was not distinctly marked until the 16th Olympiad, or 719 years before the commencement of our era. At this time, Candaules, king of Lydia, purchased a picture called the Battle of the Magnetes, for which he paid its weight in gold, although painted on boards. The name of the fortunate artist was Bularchus.

25. Notwithstanding the fame of this picture, Aglaophon and Polygnotus, of Thasos, who flourished 300 years after this period, were the first eminent painters. Polygnotus is said to have been the first who gave a pleasing air to the draperies and headdresses of females, and to have opened the mouth so far as to exhibit the beauty of the teeth.

26. Still, painting is considered to have been in an inferior state, until the appearance of Timanthes, Parrhasius, and Zeuxis, who flourished about 375 years before Christ. These again were surpassed by their successors, Protogenes, Pamphilus, Melanthius. Antiphilus, Theon, Euphranor, Apelles, and Aristides, who carried the art to the greatest perfection to which it attained in ancient times.

27. Of the preceding list of artists, Apelles was the most famous, especially as a portrait painter. He was the intimate friend of Alexander the Great, who

would never permit any other person to paint his likeness. His most celebrated painting, was this prince holding the lightning with which the picture is chiefly illuminated. By a happy application of perspective and chiaro-scuro, the hand with the lightning seemed to project from the picture.

28. From the time of these great masters, painting gradually declined, although the art continued to be practised by a succession of eminent men, who contended against the blighting influence of the luxury and the internal broils of their countrymen. But soon after Greece became subject to the Roman power, the practice of the fine arts nearly ceased in that country.

29. Before the foundation of Rome, the arts were cultivated, to some extent, in Etruria and Calabria; but the first Roman painter mentioned in history, was Fabius, a noble patrician, who painted, in the year of the city 450, the temple of the goddess Salus, and thereby obtained for himself and family the surname of *Pictor*. Yet the citizens do not seem to have profited by this example; for no other painter appeared among them until 150 years after that period. At this time, Pacuvius, the poet, amused himself, in the decline of life, with painting the temple of Hercules.

30. They were thus inattentive to the cultivation of this, as well as of the other fine arts, because they considered warfare, and the arts which tended directly to support this interest, as alone worthy of the attention of a citizen of their republic; and painting, even after the time of Pacuvius, was considered effeminate and disgraceful. Rome, therefore, cannot be said, at any time, to have produced a single artist who could approach the excellences of those of its refined neighbors, the Greeks.

31. They, however, having ornamented their me-

tropolis and villas with specimens of the arts plundered from the cities of Greece and Sicily, began, at length, to appreciate their excellences; and finally, under the first emperors, they encouraged, with great munificence, the Greeks who resorted to their city for employment.

32. But, both sculpture and painting, as well as architecture, declined with Roman civilization. Still, they continued to exist, especially in the Byzantine or Eastern empire, although in a very inferior state. The art under consideration was preserved chiefly by its application to the purposes of Christianity. It was revived in Italy, in the beginning of the twelfth century, by means of several Grecian artists, who had been employed to ornament the churches, and other edifices at Pisa, Venice, and Florence.

33. The works of Apollonius, one of these Greeks, excited in Giovanni Cimabue a spirit of emulation; and, having been initiated into the practice of the art, he executed a picture of the Virgin Mary, as large as life, for a church dedicated to her, at Florence. This production excited enthusiastic delight in his fellow-citizens, who carried it in procession, with the sound of trumpets, to its place of destination, and celebrated the day as a public feast.

34. Encouraged by this applause, Cimabue pursued the art with ardor; and, although considered a prodigy in his time, his utmost efforts failed to produce tolerable specimens of the art. He, however, far excelled his immediate predecessors; and, by introducing more correct proportions, by giving more life and expression to his figures, and by some other improvements, he became the founder of the art as it exists in modern times. He was born at Florence, in 1240, and died at the age of sixty.

35. The favorite pupil of Cimabue, was Giotto, whom he raised from a shepherd to be a painter; and

by him the art was still more relieved from the Greek imperfections. He abandoned the use of labels as means of distinguishing the different figures of a picture, and aimed at, and attained to, real expression. He marked out to the Italians the course in which the art should be pursued, as Polygnotus had done to the Greeks near 1800 years before; although, lik him, he failed in fully exemplifying his principles.

36. His abilities procured him the patronage of Pope Boniface VIII., who employed him at Rome. From this time, the art of painting became attached to the papal dignity, and few succeeding pontiffs have neglected its use. The skill and celebrity of this ingenious artist excited great emulation, and the arts having obtained an earnest of profit and honor, no longer wanted skilful professors or illustrious patrons.

37. In 1350, fourteen years after the death of Giotto, his disciple, Jacopo Cassentino, and nine other artists, founded the Academy of St. Luke, at Florence. This was a grand epoch of the arts; as from this institution arose a large display of talent, increasing in splendor until, within 150 years, it gave to the world, Masaccio, Leonardo da Vinci, Michael Angelo Buonarotti, and Raphael, besides others of great ability.

38. The art advanced but little after the time of Giotto, until the appearance of Masaccio. Under the hand of this great master, painting is said to have been greatly improved; and it was to him, that the artists who succeeded were indebted for a more sure and full direction of the course in which they ought to proceed. He was born in 1402, and died in 1443.

39. Leonardo da Vinci, who was born about two years after the death of Masaccio, brought the art to still greater perfection; and being endowed with uncommon genius, all the arts and sciences did not seem to afford a field sufficient for the exertion of his tal-

ents. He grasped at all, and succeeded far better than his predecessors in everything he undertook; but he wasted much of his time in experiments. Had he confined his great powers to the art of painting, he would probably have never been exceeded.

40. About the year 1410, oil came to be used as a vehicle for paints. It seems to have been first applied to this purpose in Flanders, by John Van Eyck, of Brussels; or it was, at least, first used by him successfully. The first hint of its utility in this application is thought, with reason, to have been obtained from its use as a varnish to pictures painted in water-colors.

41. The art of painting was introduced into Flanders about the time of Giotto, by several Flemings, who had been to Italy for the express purpose of learning it. It was also diffused in practice, about the same time, in Germany; and a particular style of the art grew up in each of these countries. But it was in Italy alone that the art may be said to have flourished in a high state of cultivation; and even there, the principal productions originated from artists of the Florentine school.

42. The art of painting was perfected, perhaps, as far as human ability can carry it, in the first half of the sixteenth century, by Michael Angelo Buonarotti, Raphael, Titian, and Correggio; although it cannot be said that all its excellences were united in the productions of any one of these distinguished professors. Such a union has never yet been displayed, nor can it hardly be expected.

43. The art was essentially aided in its progressive stages of advancement by the liberal patronage of the family of the Medici, at Florence, and by the pontiffs, at Rome. Angelo and Raphael were both employed at Rome by Julius II. and Leo X., as well as by others who succeeded them in the papal chair,

in ornamenting the palaces and sacred buildings. Their productions have never been exceeded in any country, and they still remain the objects of careful study by artists of this profession.

44. Titian was also liberally patronised at Rome, and in other parts of Italy, as well as in Spain and Germany, chiefly as a portrait and landscape painter. The unrivalled productions of these great masters, however, were fatal to the art in Italy, since their superior excellence extinguished emulation, by destroying the prospect of equal or superior success.

45. The flourishing state of the art in Italy, for so long a period, might be expected to have produced a taste for its cultivation in other parts of Europe; but this was the case only to a limited extent. No other countries have yet been particularly distinguished for artists in this branch of the fine arts, except Flanders and Holland; and these were chiefly indebted for the distinction to Peter Paul Rubens, of Antwerp, who was born at Cologne, in 1577, and to Paul Van Rhyn Rembrandt, who was born in 1606, in his father's mill, near Leyden. Some of the scholars of these masters were eminent painters. Anthony Vandyck, a pupil of the former, in particular, is said to have never yet been equalled as a portrait-painter.

46. Very little is known of the art in Spain, until about the year 1500, although it is supposed to have been cultivated with some success before that time. The examples which were left there by Titian produced a favorable impression, and several native artists of considerable eminence afterwards appeared; but the art became nearly extinct in the following age.

47. The proximity of France to Italy, and the employment of Leonardo da Vinci and other eminent artists of Italy by Francis I., together with the establishment of a school of fine arts, as stated in the pre-

ceding article, might have been expected to lay the foundation of exalted taste in this kingdom. Nevertheless, the only French painters whose names have come down to us with any pretensions to excellence for one hundred and fifty years, were Jean Cousin, Jaques Blanchard, Nicholas Poussin, and Charles Le Brun. The last, although inferior to Poussin, is at the head of the French school of painting.

48. The successors of Le Brun were not wanting in ability, yet, with a few exceptions, they failed in reaching an enviable eminence in the art, on account of their servile imitation of the false taste of their popular model. The fantastic style of Le Brun became unpopular in France some time previous to the revolution in that country; and another, of an opposite character, and by artists of other nations thought to be equally distant from true taste, has been since adopted.

49. Very little is known of the state of the fine arts in England until the time of Henry VIII., who encouraged the abilities of Hans Holbein, an eminent painter from Switzerland. But painting and sculpture, and particularly the former, having become intimately interwoven with the religion of the Church of Rome, fell into disrepute in England after the change of opinion on this subject in that country. They, however, began to revive in the eighteenth century, and England and English America have since produced some eminent painters, among whom are Hogarth, Reynolds, Opie, West, Copley, Trumbull, and Peale.

THE ENGRAVER.

ENGRAVING is the art of cutting letters or figures in wood, metals, or stone. It was practised in very ancient times, and in different countries, for the purposes of ornament and monumental inscription; but the idea of taking impressions on paper, or on any other substance, from engraved surfaces, is comparatively modern.

THE WOOD ENGRAVER.

1. The Chinese are said to have been the first who engraved figures or letters on wood, for the purpose of printing. The precise time at which they commenced the practice, is totally unknown; but a book printed by them in the tenth century, is now extant. It is thought by some antiquarians, that the Euro-

peans derived the art from the Chinese, through the Venitians, who traded in that part of the world earlier than any other Europeans.

2. This opinion is somewhat probable, from the circumstance that the tools employed by the early engravers in Europe, are similar to those used in China; and also, like the Chinese, they engraved on the side of the grain. However this may be, it is certain that the art was practised in various parts of Europe in the fourteenth century. The earliest subjects executed, were figures of saints, rudely engraved in outline. The prints taken from them were gaily colored, and sold to the common people as original paintings. The principal persons engaged in this traffic were monks, to whom the art was confined for a considerable time.

3. At length, larger subjects, with inscriptions in imitation of manuscript, were executed. The success of these prints gave rise to a more extensive application of the art. Scriptural subjects, of many figures, with texts of scripture, were engraved, and impressions were taken from them on one side of the paper, two sheets being pasted together to form a leaf. Entire sets were bound up together, and thus were formed the first printed books, which, being produced entirely from wood-cuts, are known by the name of *block-books*. These books made their appearance about the year 1420.

4. One of the earliest of these productions is denominated "The Apocalypse of St. John;" another, "The Poor Man's Bible." But one of the latest and most celebrated, is called "The Mirror of Salvation," published in 1440. Part of the text was printed from solid blocks, and part, from moveable wooden types. From this fact, it is easy to discover the origin of printing. After this, most, if not all, of the books, were printed from moveable types; but, as they were

embellished with wood cuts, the demand for such engravings was very much increased, although they were, at first, by no means elegant.

5. Near the close of the fifteenth century, the art began to assume a higher character, principally by the talents of Michael Wolgermuth and William Pluydenwurf. Albert Durer made still greater improvements, and, in 1498, published his celebrated Apocalypse of St. John, printed from folio blocks. Other celebrated engravers succeeded him in the sixteenth century, which may be considered the era when wood engraving was at its highest point of elevation. After this, the art declined, and was considered of little importance, until it was revived in 1775, by the distinguished William Bewick, of Newcastle, England. It is still practised, especially in England and the United States, in a manner which reflects credit on the ingenuity of the age.

6. The earlier artists operated on various kinds of wood, such as the apple, pear, and beech; but these, being too soft, are now used only for calico-printing and other common purposes. Box-wood, on account of its superior texture, is used for every subject that can be termed a work of art. That from Turkey is the best.

7. The engravers, in the infancy of the art, prepared the wood as the common block-cutters now do. The tree was cut the way of the grain, in planks, and of course they engraved on the side of the grain, as upon a board. This mode of preparation enabled them to execute larger subjects. The engravers now prefer the end of the grain, and therefore cut the log transversely.

8. The end on which the engraver is to exert his skill, is planed and scraped, to render the surface smooth, and the block having been cut to the proper size, the drawing is made upon it in India ink, or with

a lead-pencil. The block is now ready for the artist who, in executing the work, holds it with one hand, on a cushion made of sand and leather, while, with the other, he cuts away the superfluous wood. The part intended to make the impression in printing, is left standing.

9. Wood engravings, well executed, are scarcely inferior to those of copper and steel, and, for many purposes, they are preferred. They are remarkably convenient, since they can be inserted into a page of types, where illustrations or embellishments may be required, and be printed without separate expense. They will also bear a great number of impressions— generally 100,000. In this respect, they are decidedly superior to metallic plates. They can likewise be multiplied indefinitely by the process of stereotyping.

THE COPPERPLATE ENGRAVER.

1. THE engravers on metallic surfaces are termed copperplate engravers, not because copper is the only metal on which they exert their skill, but because it is the one on which they usually operate. The plates are prepared for the artist by the coppersmith, by rubbing them with brickdust and charcoal, after having cut them of a proper size from sheets of copper.

2. The instruments employed by this artist are few and simple, the principal of which are, the *graver*, the *dry-point*, the *scraper*, and the *burnisher*. The *graver* is a small bar of steel, of a square or lozenge form, and, with the short handle into which it is inserted, about five inches in length. One of the angles of the bar is always on the under side of the instrument, and the point is formed by bevelling the end from the upper side, or angle. The square form is used for broad strokes, and the lozenge for fine ones.

3. The *dry-point*, or needle, is a steel wire with a

long cylindrical handle; or it is simply a wire of sufficient length and size to be used without a handle. The *scraper* has nearly the form of a triangular pyramid; and the cutting part, which has three edges, is two or three inches long. The *burnisher* has a form nearly conical, and, without the handle, is about three inches long. The last two instruments are frequently made of the same piece of steel, properly forged at each end. In such case, the middle part of the steel is the handle by which they are held.

4. Of engraving on copper, the following are the principal varieties or styles: 1. Line engraving; 2. Stippling; 3. Etching; 4. Mezzotinto; 5. Aquatinta. For the purpose of conveying some idea of these different branches, we will describe them under distinct heads.

5. *Line engraving.* The first thing done, in this species of engraving, is to transfer to the plate an exact copy of the outlines of the design to be executed. In accomplishing this, the plate is moderately heated, and covered with a thin coating of white wax. A piece of transparent paper is then laid over the design to be copied, and traced in outline with a black-lead pencil. The outline thus sketched is turned down upon the coating of white wax, and the whole is subjected to the action of a rolling-press; or it is kept for a while under heavy weights. By the application of this pressure, the lines are transferred from the paper to the wax on the plate in a reversed position, which is necessary to make the impression of the finished plate resemble the original.

6. The pencil-marks on the wax having been lightly traced on the copper with the dry-point, and the wax having been melted off, a perfect outline is found on the plate. Small subordinate parts of the design are transferred to the plate in the same manner, except that the transparent paper is brought in forcible

contact with the waxed surface by means of the burnisher.

7. At this stage of the process, the artist commences the use of the graver. While operating with this instrument, he holds the handle in the palm of his hand, and pushes the point forward with a firm and steady motion, until a line is produced by a removal of a portion of the metal. By a succession of such strokes, judiciously applied, the work is completed. The *burrs*, or little elevations of the copper, left by the graver on each side of the lines, are removed by means of the scraper and burnisher. Mistakes or blemishes are erased from the plate, either with the burnisher, or by friction with charcoal.

8. *Stippling.* The second mode of engraving is called stippling. This resembles the last method in its process, except that the effect is produced by means of minute punctures or excavations, instead of lines. These are made either with the dry-point or graver. When produced by the former instrument, they are of a circular form; when by the latter, they are rhomboidal or triangular. This style of work is always more slow, and consequently more expensive, than engraving in lines. It has, however, some advantages in the softness and delicacy of its lights and shades, and the prints struck from it approach more nearly to paintings.

9. *Etching.* This mode of engraving is far more easy than any other, being performed chiefly by chemical corrosion. In fact, any person who can draw, may *etch* coarse designs tolerably well, after having learned the theory of the operation. To perform it, the plate is first covered with a thin coating of some resinous substance, upon which the acid employed can have no action. The design, and all the lines it requires, are next traced on the plate with steel points, called *etching needles*, which are instruments similar to the dry-point.

10. The second part of the process is the corrosion, or, as it is technically called, *biting in*. This is effected by pouring upon the design a quantity of diluted nitric acid, after having surrounded the edges of the plate with a wall of soft wax, to prevent the escape of the fluid. A chemical action immediately takes place in all the lines or points where the copper has been denuded by the needle. After the first biting has been continued long enough, in the judgment of the operator, the acid is poured off, and the plate examined.

11. The light shades, if found sufficiently deep, are then covered with varnish, to protect them from further corrosion. The biting is then continued for the second shades, in the same manner, and afterwards, for the third and succeeding shades, until the piece shall have been finished. The plate having been cleaned, and carefully examined by the aid of a proof impression, the deficiencies which may be discovered are supplied with the graver.

12. *Mezzotinto*. In the production of this kind of engraving, the whole surface of the plate is first roughened, or covered with minute prominences and excavations too small to be obvious to the naked eye; so that an impression taken from it, in this state, would present a uniform velvety, black appearance. This roughness is produced mechanically by means of a small toothed instrument, called a *cradle*.

13. When the plate has been thus prepared, the rest of the process is comparatively easy. It consists in pressing down or rubbing out the roughness of certain parts of the plate, with the burnisher and scraper. Where strong lights are required, the plate is restored to a smooth surface; for a medium light, it is moderately burnished, or partially erased; and, for the deepest shades, the ground is left entire, and sometimes etched, and corroded with nitric acid.

Impressions from mezzotinto plates approach more nearly to oil paintings than any other prints. This kind of engraving was invented by Prince Rupert, in 1649.

14. *Aqua-tinta.* There are several methods by which this kind of engraving can be executed; we, however, will describe the one which seems to be the most simple and obvious. The outline of the picture having been etched or engraved in the usual manner, the surface of the copper is sprinkled equally with minute particles of rosin. This dust is fixed to the surface by heating the plate until the rosin has melted.

15. The ground having been thus laid, the parts of the plates not intended to be occupied by the design are *stopped out* by means of thick varnish. The plate is now surrounded with a wall of wax, as for etching, and diluted nitric acid is poured upon it. A chemical action immediately takes place, by which the surface exposed between the resinous particles is minutely excavated.

16. The lighter shades are stopped out at an early stage of the process, and the *biting in* is continued for the darker ones. After the plate is judged to be sufficiently corroded, it is cleansed, and an impression is taken on paper. The process is finished by burnishing the shades, to give them greater softness, and by touching up the defective parts with the graver.

17. This mode of engraving is well adapted to light subjects, sketches, landscapes, &c.; but, owing to the fineness of the ground, the plates wear out rapidly, and seldom yield, when of ordinary strength, more than six hundred impressions. The prints taken from such plates bear a strong resemblance to paintings in Indian ink, or to drawings in black-lead pencil. Aqua-tinta is the most precarious kind of engraving, and requires much attention on the part of the artist. It was invented by a Frenchman, named Leprince.

who, for a time, kept the process a secret, and sold his impressions for original drawings.

18. *Steel engraving.* The process of engraving on steel plates differs but little in its details from that on copper plates; and the chief advantage derived from this method, arises from the hardness or toughness of the material, which renders it capable of yielding a greater number of impressions.

19. This mode of engraving was first practised, in England, by the calico-printers; but steel was first employed for bank-notes, and for common designs, by Jacob Perkins, of Newburyport, Massachusetts; and by him, in conjunction with Asa Spencer, of New-London, and Gideon Fairman, of Philadelphia, the use of steel in this application was generally introduced, not only in the United States, but also in Great Britain, some time before the year 1820.

20. The plates are prepared for the engraver from sheets of steel about one-sixth of an inch in thickness. A plate cut from a sheet of this kind is first softened by heating it with charcoal, and suffering it to cool gradually in the atmosphere. It is next *planished*, or hammered on a peculiar kind of anvil, to make it perfectly level, and afterwards ground on one side upon a grindstone. The operation is completed by polishing it with Scotch stone and charcoal. When steel was first substituted for copper, it was hardened before it was used in printing; but it is now used in its soft state, as it comes from the hands of the artist.

THE COPPERPLATE-PRINTER.

1. The copperplate-printer takes impressions on paper from engraved plates by means of a rolling press. This machine, together with some of the operations in its application, are well exhibited in the above picture.

2. The period at which the practice of printing rom engraved plates commenced, cannot be ascertained with any degree of certainty. The Dutch, the Germans, and the Italians, contend for the honor of introducing it; but the weight of testimony seems to be in favor of the claims of the Italian sculptor and goldsmith, Tommaso Finiguera, who flourished at Florence, about the middle of the fifteenth century.

3. It is stated that this artist, accidentally spilling some melted brimstone on an engraved plate, found,

on its removal, an exact impression of the engraving, marked with black, taken out of the strokes. This suggested to him the idea of taking an impression in ink on paper, by the aid of a roller. It is hardly necessary to state, that the experiment succeeded. Copperplate-printing was not used in England until about 150 years after its first employment at Florence, when it was introduced from Antwerp, by Speed.

4. The ink used in this kind of printing is made of a carbonaceous substance, called Frankfort black, and linseed or nut oil. Oil is used, instead of water, that the ink may not dry during the process; and it is boiled till it has become thick and viscid, that it may not spread on the paper. The materials are incorporated and prepared with the stone and muller, as painters prepare their colors.

5. In taking impressions from an engraved plate, it is first placed on an iron frame over a heated stove, or over a charcoal fire in a furnace, and while in this position, the ink is spread over it with a roller covered with coarse cloth, or with a ball of rubber made of the same material, and faced with buckskin. The heat renders the ink so thin that it can penetrate the minute excavations of the engraving. The plate having been thus sufficiently charged, is wiped first with a rag, then with the hand, until the ink has been removed from every portion of it, except from the lines of the engraving.

6. The plate is next placed on the platform of the press, with its face upwards, and the paper, which has been previously dampened, is laid upon it. A turn of the cylinders, by means of the arms of the cross, carries the plate under a strong pressure, by which portions of the paper are forced into all the cavities of the engraving. The ink, or part of it, leaves the plate, and adheres to the paper, giving an exact representation of the whole work of the artist. The roll-

er by which the pressure is applied is covered with several thicknesses of broadcloth.

7. The number of good impressions yielded by engraved copperplates, depends upon various circumstances, but chiefly on the fineness and depth of the work; and these qualities depend mainly upon the sty.e in which it has been executed. Line engravings will admit of four or five thousand, and, after having been retouched, a considerable number more.

8. Plates of steel will yield near ten times as many good impressions as those of copper, and this too without being hardened. Besides, an engraving on steel may be transferred to a softened steel cylinder, in such a manner that the lines may stand in relief; and this cylinder, after having been hardened, may be brought in forcible contact with another plate, and thus the design may be multiplied at pleasure.

9. The bank-note engravers have now a great variety of designs and figures on steel rollers, which they can easily transfer to new plates. This practice, as applied to plates for bank-notes, originated with Jacob Perkins. It is supposed that he must have been led to it by an English engraver in his employ, who may have explained to him the manner in which the British calico-printers produced engravings on copper cylinders. This is not altogether improbable, since the principle in both cases is substantially the same.

10. In consequence of the increased demand for maps and pictorial embellishments in books, as well as for single prints as ornaments for rooms, engraving and copperplate-printing have become employments of considerable importance; and these arts must doubtless continue to flourish to an indefinite extent, in a country where the taste for the fine arts is rapidly improving, and where wealth affords the means of liberal patronage.

THE LITHOGRAPHER.

1. The word *lithography* is derived from two Greek words—*lithos*, a stone, and *grapho*, to write; and the art to which the term is applied has reference to the execution of letters, figures, and drawings, on stone, and taking from them fac-simile impressions. The art is founded on the property which stone possesses, of imbibing fluids by capillary attraction, and on the chemical repulsion which oil and water have for each other.

2. Every kind of calcareous stone is capable of being used for lithography. Those, however, which are of a compact, fine, and equal grain, are best adapted to the purpose. The quarries of Solenhofen, near Pappenheim, in Bavaria, furnished the first plates, and none have yet been found in any other

place, to equal them in quality; although some that answer the purpose tolerably well, have been taken from quarries in France and England.

3. In preparing the stones for use, they are first ground to a level surface, by rubbing two of them face to face, sand and water being interposed. Then, if they are designed for *ink drawings*, they are polished with pumice-stone; but, if for *chalk drawings*, with fine sand, which produces a grained surface adapted to holding the chalk.

4. When stones of proper size and texture cannot be conveniently obtained, slabs are sometimes constructed of lime and sand, and united with the caseous part of milk. The first part of the process which may be considered as belonging peculiarly to the art, consists in making the drawing on the stone. This is done either in ink, with steel pens and camel's hair pencils, or with crayons made of lithographic chalk. The process of drawing on stone differs but little from that on paper, with similar means.

5. For lithographic ink, a great number of receipts have been given; but the most approved composition consists of equal parts of wax, tallow, shell-lac, and common soap, with a small proportion of lamp-black. Lithographic chalk is usually composed of the same materials, combined in different proportions.

6. When the drawing has been finished, the lithographic printer prepares it for giving impressions, by using upon its surface a weak solution of acid and other ingredients, which corrode the surface of the stone, except where it is defended from its action by the grease of the chalk or ink. As soon as the stone has been sufficiently eaten away, the solution is removed by the application of spirits of turpentine and water.

7. The ink employed in this kind of printing, is similar in its composition to other kinds of printing

ink. It is applied to the drawing by means of a small wooden cylinder covered with leather. The paper, which has been suitably dampened, is laid upon the stone, and after it has been covered, by turning down upon it a thick piece of leather stretched upon an iron frame, a crank is turned which brings the stone successively under the press.

8. An impression of the drawing having been thus communicated to the paper, the sheet is removed, and the process is repeated, until the proposed number of prints have been taken. Before each application of the ink, the whole face of the stone is moderately wet with water by means of a sponge; and although the roller passes over the whole surface of the stone, yet the ink adheres to no part of it, except to that which is covered with the drawing.

9. The number of impressions which may be taken from chalk drawings, varies according to their fineness. A fine drawing will give fifteen hundred; a coarse one, twice that number. Ink drawings and writings give considerably more than copperplates, the finest yielding six or eight thousand, and strong lines and writings many more.

10. Impressions from engravings can be multiplied indefinitely, with very little trouble, in the following manner. A print is taken in the usual way from the engraved plate, and immediately laid with its face upon water. When sufficiently wet, it is carefully applied to the face of a stone, and pressed down upon it by the application of a roller, until the ink is transferred to the stone. Impressions are then taken in the manner before described.

11. The invention of lithography is ascribed to Aloys Senifelder, the son of a performer at the theatre of Munich. Having become an author, and being too poor to publish his works in the usual way, he tried many plans, with copperplates and compositions.

in order to be his own printer. A trial on stone, which had been accidentally suggested, succeeded. His first essays to print for publication, were some pieces of music, executed in 1796.

12. The first productions of the art were rude, and of little promise; but, since 1806, its progress has been so rapid, that it now gives employment to a great number of artists; and works are produced, which rival the finest engravings, and even surpass them in the expression of certain subjects. The earliest date of the art in the United States, is 1826, when a press was established at Boston, by William Pendleton.

THE AUTHOR.

1. THE word author, in a general sense, is used to express the originator or efficient cause of a thing; but, in the restricted sense in which it is applied in this article, it signifies the first writer of a book, or a writer in general. The indispensable qualifications to make a writer are—a talent for literary composition, an accurate knowledge of language, and an acquaintance with the subject to be treated.

2. Very few persons are educated with the view to their becoming authors. They generally write on subjects pertaining to the profession or business in which they have been practically engaged: a clergyman writes on divinity; a physician, on medicine; a lawyer, on jurisprudence; a teacher, on education; and a mechanic, on his particular trade. There are

subjects, however, which occupy common ground, on which individuals of various professions often write.

3. Authorship is founded upon the invention of letters, and the art of combining them into words. In the earliest ages of the world, the increase of knowledge was opposed by many formidable obstacles. Tradition was the first means of transmitting information to posterity; and this, depending upon the memory and will of individuals, was exceedingly precarious.

4. The chief adventitious aids in the perpetuation of the memory of facts by tradition, were the erection of monuments, the periodical celebration of days or years, the use of poetry, and, finally, symbolical drawings and hieroglyphical sketches. Nevertheless, history must have remained uncertain and fabulous, and science in a state of perpetual infancy, had it not been for the invention of written characters.

5. The credit of the invention of letters was claimed by the Egyptians, Phœnicians, and Jews, as well as by some other nations; but as their origin preceded all authentic history not inspired, and as the book of inspiration is silent in regard to it, no satisfactory conclusion can be formed on this point. Some antiquarians are of opinion, that the strongest claims are presented by the Phœnicians.

6. The Pentateuch embraces the earliest specimen of phonetic or alphabetic writing now extant, and this was written about 1500 years before Christ. Many persons suppose that, as the Deity himself inscribed the ten commandments on the two tables of stone, he taught Moses the use of letters; and, on this supposition, is founded the claim of the Jewish nation to the honor of the first human application of them.

7. If we may believe Pliny, sixteen characters of the alphabet were introduced into Greece by Cadmus, the Phœnician. in the days of Moses; four more were

added by Palamedes during the Trojan war, and four afterwards, by Simonides. Alphabetical writing evidently sprung from successive improvements in the hieroglyphical system, since a great part of the latter has been lately discovered to be syllabic or alphabetic.

8. A considerable number of very ancient alphabets still exist on the monumental remains of some of the first post-diluvian cities, and several of later date, in manuscripts which have descended to our times. The letters employed in different languages have ever been subject to great changes in their conformation. This was especially the case before the introduction of the art of printing, which has contributed greatly towards permanency in this respect.

9. The mode of arranging the letters in writing has, also, varied considerably. Some nations have written in perpendicular lines, as the Chinese and ancient Egyptians; others from right to left, as the Jews; and others, again, alternately from left to right, as was the method at one period among the Greeks. The mode of writing from left to right now generally practised, is preferable to any other, since it leaves uncovered that portion of the page upon which writing has been made.

10. In ancient times, literary productions were considered public property; and, consequently, as soon as a work was published, transcribers assumed the right to multiply copies at pleasure, without making the authors the least remuneration. They, however, were sometimes rewarded with great liberality, by princes or wealthy patrons. This literary piracy continued, until a long time after the introduction of the art of printing.

11. In almost every kingdom of Europe, and in the United States, the exclusive right of authors to publish their own productions, is now secured to them by law, at least for a specified number of years. The

first legislative proceeding on this subject in England, took place in 1662, when the publication of any book was prohibited, except through the permission of the lord-chamberlain. The title of the book, and the name of the proprietor, were, also, required to be entered in the record of the Stationers' Company.

12. This and some subsequent acts having been repealed in 1691, literary property was left to the protection of the common law, by which the amount of damages which could be proved to have actually occurred in case of infringement, could be recovered, and no more. New applications were, therefore, made to parliament; and, in 1709, a statute was passed, by which the property of copyright was guarded for fourteen years, with severe penalties. This privilege was connected with the condition, that a copy of the work be deposited in nine public libraries specified in the act.

13. In 1774, the Parliament decided that, at the end of fourteen years, the copyright might be renewed, in case the author were still living. The law continued on this footing until 1814, when the contingency with regard to the last fourteen years was removed; and, if the author still survived, the privilege of publication was extended to the close of his life.

14. In the United States, the jurisdiction of this subject is vested by the Constitution in the Federal Government; and, in 1790, a law was passed by Congress, securing to the authors of books, charts, maps, engravings, &c., being citizens of the United States or resident therein, privileges like those granted in England, in 1774. In 1831, the law was altered, and again made to conform to that of England in regard to the period of the privileges. The English and American laws differ in no essential provision. Until the year 1839, foreigners were permitted to hold copyrights in England.

15. In France, the first statute regarding literary property was passed in 1793, when the right of authors to their works was secured to them during their lives, and to their heirs for ten years after their decease. The decree of 1810 extended the right of the heirs to twenty years. In Russia, the period of copyright is the same as in France, and the property is not liable for the payment of the author's debts.

16. In some of the German states, the right is given for the lifetime of the author; in others, it is made perpetual, like any other property; but then the work may be printed with impunity in any of the other states in which a right has not been secured. In Germany and Italy, especially, authors are very poorly remunerated; and in Spain, the book trade has been so much oppressed by a merciless censorship, that authors are compelled to publish their works on their own account.

17. From the preceding statement it appears, that few legislators have been willing to place the productions of intellectual labor on the same honorable footing with other kinds of property. No reason, however, can be assigned for the distinction, except the unjust and piratical usage of two or three thousand years.

18. Authors seldom publish their own works. They generally find it expedient, and, in fact, necessary, to intrust this part of the business to booksellers and publishers, from whom they usually receive a specified amount for the entire copyright, or a certain sum for each and every copy which may be sold during the term of years which may be agreed upon. The compensation is commonly insufficient to pay them for preparing the works for the press; but they are as well paid in this country as in any other. In this particular, however, there has been a manifest improvement within the last ten years.

THE PRINTER.

1. From what has been said in a preceding article, it is manifest that the art of printing arose from the practice of engraving on wood. Letters were cut on wood as inscriptions to pictures, and were printed at the same time with them, by means of a hand-roller. The impressions were taken on one side of the paper; and, in order to hide the nakedness of the blank side, two leaves were pasted together. These leaves were put up in pamphlet form, and are now known under the denomination of *block-books*, because they were printed from wooden blocks.

2. Although the art of typographical printing can be clearly traced to wood engraving, yet so much uncertainty rests upon its history, that the honor of its invention is claimed by three cities—Harlem, in Hol-

land, and Strasburg and Mentz, in Germany; and, at the present time, it is difficult to determine satisfactorily the merits of their respective claims. The obscurity on this point has arisen from the desire of the first printers to conceal the process of the art, that their productions might pass for manuscripts, and that they might enjoy the full benefit of their invention.

3. The advocates of the claims of Harlem state, that Laurentius Coster applied wooden types, and some say, even metal types, as early as 1428, and that several persons were employed by him in the business up to the year 1440, when his materials were stolen from him by one of his workmen or servants, named John, while the family were engaged in celebrating the festival of Christmas eve. The thief is said to have fled first to Amsterdam, then to Cologne, and, finally, to have settled in Mentz, where, within a twelvemonth, he published two small works, by means of the types which Laurentius Coster had used.

4. These claims in favor of Harlem, however, were not set forth until 120 years after the death of Coster; and the whole story, as then stated by Hadriamus Junius, was founded altogether upon traditionary testimony. Perhaps wood engravings, with inscriptions, may have been executed there; if so, the account may have originated from that circumstance.

5. The statements which seem to be the most worthy of credit, bestow the honor of this invention on a citizen of Mentz. Here, it appears, that John Geinsfleisch, or Guttemburg senior, published two small works for schools, in 1442, on wooden types; but, not having the funds necessary to carry on the business, he applied to John Faust, a rich goldsmith, who became a partner, in 1443, and advanced the requisite means. Soon afterwards, J. Meidenbachius and some others were admitted as partners.

6. In the following year, John Guttemburg, the

brother of Geinsfleisch, made an addition to the firm. For several years before this union, or from 1436, Guttemburg had been attempting to complete the invention at Strasburg; but it is said that he had never been able to produce a clean printed sheet. The brothers may, or may not, have pursued their experiments without receiving any hints from each other, before their union at Mentz.

7. Soon after the formation of this partnership, the two brothers commenced cutting *metal types*, for the purpose of printing an edition of the Bible, which was published in Latin, about the year 1450. Before this great achievement of the art had been effected, Geinsfleisch appears to have retired from the concern, some say, on account of blindness.

8. The partnership before mentioned, was dissolved, in 1450, and Faust and Guttemburg entered into a new arrangement, the former supplying money, the latter, personal services, for their mutual benefit; but various difficulties having arisen, this partnership was also dissolved, in 1455, after a lawsuit between them, which was decided against Guttemburg.

9. Faust, having obtained possession of the printing materials, entered into partnership with Peter Shœffer, who had been for a long time a servant, or workman, in the printing establishment. In 1457, they published an edition of the *Psalter*, which was then considered uncommonly elegant. This book was, in a great measure, the work of Guttemburg, since, during the four years in which it was in the press, he was, for two years and a half, the chief operator in the printing-office.

10. Guttemburg, by the pecuniary aid of Conrad Humery and others, established another press in Mentz, and, in 1460, published the " *Catholicon Joannis Januensis.*" It was a very handsome work, but not equal in beauty to the Psalter of Faust and Shœf-

fer. The latter was the first printed book known to have a genuine date. From this time, it has been the practice for printers to claim their own productions, by prefixing to them their names.

11. Notwithstanding the great advancement which had been made in the art of printing, the invention cannot, by any means, be considered complete, until about the year 1458, when Peter Shœffer contrived a method of casting types in a matrix, or mould. The first book executed with cast metal types was called " *Durandi Rationale Divinorum Officiorum*," published in 1459. Only the smaller letters, however, were of this description, all the larger characters which occur, being *cut types*. These continued to be used, more or less, as late as the year 1490.

12. In 1462, Faust carried to Paris a number of Latin Bibles, which he and Shœffer had printed, and disposed of many of them as manuscripts. At first, he sold them at five or six hundred crowns, the sums usually obtained by the scribes. He afterwards lowered the price to sixty. This created universal astonishment; but, when he produced them according to the demand, and when he had reduced the price to thirty, all Paris became agitated.

13. The uniformity of the copies increased the wonder of the Parisians, and information was finally given against him to the police as a magician. He was accordingly arrested, and a great number of his Bibles were seized. The red ink with which they were embellished, was supposed to be his blood. It was seriously adjudged, that the prisoner was joined in league with the devil; and had he not disclosed the secret of his art, he would probably have shared the fate of those whom the magistrates of those superstitious times condemned for witchcraft.

14. It may be well to inform the reader, that, although the story of Faust's arrest, as above detailed,

is related as a fact by several authors, yet by others it is thought to be unworthy of credit. It is also generally supposed, that the celebrated romance of " Doctor Faustus and the devil" originated in the malice of the monks towards Faust, whose employment of printing deprived them of their gain as copiers. It seems more probable, however, that it arose from the astonishing performances of Doctor John Faust, a dealer in the black art, who lived in Germany in the beginning of the sixteenth century.

15. Faust and Shœffer continued their printing operations together, at least, until 1466, about which time it is conjectured, that the former died of the plague, at Paris. Geinsfleisch, or, as he is sometimes called, Guttemburg senior, died in 1462; and his brother Guttemburg junior, in 1468, after having enjoyed, for three years, the privileges of nobility, which, together with a pension, had been conferred upon him by Archbishop Adolphus, in consideration of his great services to mankind.

16. More copies of the earliest printed books were impressed on vellum than on paper; but very soon paper was used for a principal part of the edition, while a few only were printed on vellum, as curiosities, to be ornamented by the illuminators, whose ingenious art, though in vogue before and at that time, did not long survive the rapid improvements in printing.

17. We are informed, that the Mentz printers observed the utmost secrecy in their operations; and, that the art might not be divulged by the persons whom they employed, they administered to them an oath of fidelity. This appears to have been strictly adhered to, until the year 1462, when the city was taken and plundered by Archbishop Adolphus. Amid the consternation which had arisen from this event, the workmen spread themselves in different direc

tions; and, considering their oath no longer obligatory, they soon divulged the secret, which was rapidly diffused throughout Europe.

18. Some idea may be formed of the celerity with which a knowledge of printing was extended, from the fact that the art was received in two hundred and three places, prior to the year 1500. It was brought to England, in 1471, by William Caxton, a mercer of the city of London, who had spent many years in Germany and Holland. The place of the first location of his press was Westminster Abbey. The first press in North America was established at Cambridge, Massachusetts, in 1639.

19. Printed newspapers had their origin in Germany. They first appeared in Augsburg and Vienna, in 1524. They were originally without date or place of impression; nor were they published at regular periods. The first German paper with numbered sheets was printed, in 1612; and, from this time, must be dated periodical publications in that part of Europe.

20. In England, the first newspaper appeared during the reign of Elizabeth. It originated in a desire to communicate information in regard to the expected invasion by the Spanish armada, and was entitled the "English Mercury," which, by authority, was printed at London by Christopher Barker, her highness's printer, in 1588.

21. These, however, were extraordinary gazettes, not regularly published. Periodicals seem to have been first extensively used by the English, during the civil wars in the time of the Commonwealth. The number of newspapers in Great Britain and Ireland amounted, in 1829, to 325, and the sums paid to the government for stamps and duties on advertisements, amounted to about £678,000 sterling.

22. No newspaper appeared in the British colonies

of America until 1704, when the "News Letter" was issued at Boston. The first paper published in Philadelphia, was issued in 1719; the first in New-York. in 1733. In 1775, there were 37; and in 1801, there were, in the whole United States, 203; in 1810, 358; at the present time, there are about 1500, and the number is annually increasing.

23. The first periodical paper of France originated with Renaudot, a physician in Paris, who, for a long time, had been in the habit of collecting news, which he communicated verbally to his patients, with the view to their amusement. But, in 1631, he commenced the publication of a weekly sheet, called the "Gazette de France," which was continued with very little interruption, until 1827. There are now, probably, in France, about 400 periodical publications most of which have been established since the commencement of the revolution of 1792.

24. Periodicals devoted to different objects have been established in every other kingdom of Europe; but, in many cases, they are trammelled by a strict censorship of the respective governments. This is especially the case with those devoted to politics or religion. But all Europe, with its 200,000,000 of inhabitants, does not support as many regular publications as the United States, with its 17,000,000.

25. The workmen employed in a printing-office are of two kinds: *compositors*, who arrange the types according to the copy delivered to them; and *pressmen*, who apply ink on the types, and take off impressions. In many cases, and especially where the business is carried on upon a small scale, the workmen often practise both branches.

26. Before the types are applied to use, they are placed in the cells or compartments of a wooden receptacle called a *case*, each species of letter, character and space, by itself. The letters which are required

most frequently, are lodged in the largest compartments, which are located nearest to the place where the compositor stands, while arranging the types.

27. The compositor is furnished with a *composing-stick*, which is commonly an iron instrument, surrounded on three sides with ledges about half an inch in height, one of which is moveable, so that it may be adjusted to any length of line. The compositor, in the performance of his work, selects the letters from their several compartments, and arranges them in an inverted order from that in which they are to appear in the printed page.

28. At the end of each word is placed a *quadrat*, to produce a space between that and the one which follows. The quadrats are of various widths, and being considerably shorter than types, they yield no impression in printing. A thin brass rule is placed in the stick, on which each successive line of types is arranged. When the composing-stick has been filled, it is *emptied* into the *galley*, which is a flat board, partly surrounded with a rim.

29. On this galley, the lines are accumulated in long columns, which are afterwards divided into pages, and tied together with a string, to prevent the types from falling asunder, or into *pi*, as the printers term it. A sufficient number of pages having been completed to constitute a *form*, or, in other words, to fill one side of a sheet of printing-paper, they are arranged on an *imposing-stone*, and strongly locked up, or wedged together, in an iron *chase*.

30. The first impression taken from the types is called the *proof*. This is carefully read over by the author or proof-reader, or both, and the errors and corrections plainly marked in the margin. These corrections having been made by the compositor, the form is again locked up, and delivered to the pressman.

31. The pressman having dampened his paper with water, and put every part of his press in order, takes impressions in the following manner: he places the sheet upon the *tympan*, and confines it there by turning down upon it the *frisket;* he then brings them both, together with the paper, upon the form, which has been previously inked. He next turns a crank with his left hand, and thereby places the form directly under the *platen*, which is immediately brought, in a perpendicular direction, upon the types, by means of a lever pulled with his right hand.

32. After the impression has been thus communicated, the form is returned to its former position, and the printed sheet is removed. The operation just described, is repeated for each side of every sheet of the edition. In the cut at the head of this article, the pressman is represented as in the act of turning down the frisket upon the tympan. The business of the boy behind the press is to apply the ink to the types by means of the *rollers* before him. In offices where much printing is executed, the roller-boy is now dispensed with, simple machinery, attached to the crank of the press, called a *patent roller-boy*, being substituted in his place.

33. Within the present century, great improvements have been made in the printing business generally, especially in the presses, and in the means of applying the ink. In the old *Ramage* press, the power was derived from a screw which was moved by a lever; but, in those by several late inventors, from an accumulation of levers.

34. In 1814, printing by machinery was commenced in London, and rollers became necessary for inking the forms. These were made of molasses, glue, and tar, in proportions to suit the temperature of the weather. From these originated composition balls in the following year, and in 1819, hand rollers. For-

merly the ink was applied by means of pelt balls stuffed with wool.

35. The power-press first used in this country, was invented, in 1823, by Mr. Treadwell, a scientific mechanic, of Boston, who was originally a watch-maker by trade. It acts on the same principle with the hand press, and is equal to three of these of the best construction. Daniel Fanshaw, who first applied steam to printing in the United States, introduced several of these presses into New-York, in 1826. Messrs. Adams and Tufts, of Boston, have each invented a power-press which act on the same principle with Mr. Treadwell's.

36. The presses noticed in the preceding paragraph, are used chiefly in printing books and periodicals requiring moderate speed in their production. But they do not answer the purposes of the daily press in large cities, where from twenty thousand to sixty thousand impressions of a single paper are required every day. To supply this immense demand of the public was the original aim of the inventors of power-presses in England. The first attempt to construct a printing machine was made, in 1790, by William Nicholson, of London; but his machine was never brought into use. The next attempt was made by Mr. Konig, an ingenious German, who but partially succeeded. The first really useful machine was constructed by Messrs. Applegate and Cowper.

37. The machines used in this country are modifications of that originally invented by Mr. Napier, of England. The paper is brought in contact with the form of types by means of a cylinder, while the form is passing underneath it. The press is constructed with one or two cylinders. A double cylinder press will give from 4000 to 6000 impressions an hour. The improvements on this press were made by Robert Hoe & Co., who have permitted Mr. Napier to introduce them into his press in England.

THE TYPE-FOUNDER.

1. The types cast by the type-founder are oblong square pieces of metal, each having, on one end of it, a letter or character in relief. The metal of which these important instruments are composed, is commonly an alloy consisting principally of lead and antimony, in the proportion of about five parts of the former to one of the latter. This alloy melts at a low temperature, and receives and retains with accuracy the shape of the mould. Several hundred pounds of type-metal are prepared at a time, and cast into bars filled with notches, that they may be easily broken into pieces, when about to be applied to use.

2. In making types, the letter or character is first formed, by means of gravers and other tools, on the end of a steel punch. With this instrument, a *matrix*

is formed, by driving it into a piece of copper of suitable size. A punch and matrix are required for every character used in printing. A metallic mould for the body of the type is also made; and, that the workman may handle it without burning his hands, it is surrounded with a portion of wood. The mould is composed of two parts, which can be closed and separated with the greatest facility.

3. The type-metal is prepared for immediate use by melting it, as fast as it may be needed, in a small crucible, over a coal fire. The caster having placed the matrix in the bottom of the mould, commences the operation of casting by pouring the metal into the mould with a small ladle. This he performs with his right hand, while with the other he throws up the mould with a sudden jerk; then, with both hands he opens it, and throws out the type. All these movements are performed with such rapidity, that an expert hand can cast about fifty types of a common size in a minute. Some machines have been lately introduced, which operate with still greater rapidity.

4. Each type, when thrown from the mould, has attached to it a superfluous portion of metal, called a *jet,* which is afterwards broken off by hand. The jets are again cast into the pot, or crucible, and the types are carried to another room, where the two broad sides are rubbed on a grindstone. They are next arranged on flat sticks about three feet long, and delivered to the *dresser,* who scrapes the two sides not before made smooth on the grindstone, cuts a groove on the end opposite the letter, and rejects from the row the types which may be defective.

5. The whole process is completed by setting up the types in a printer's composing-stick, and tying them up with packthread. Much of the work in the type-foundry is performed by boys and females. In the preceding cut are represented a man casting types

at a furnace, and a boy breaking off the jets; also two females rubbing types on a large grindstone. The fumes arising from melted lead in the casting-room are considered deleterious to health.

6. Various sizes of the same kind of letter are extensively used, of which the following are most employed in printing books—Pica, Small Pica, Long Primer, Bourgeois, Brevier, Minion, Nonpareil, Pearl, and Diamond. A full assortment of any particular size is called a *fount,* which may consist of any amount, from five pounds to five hundred, or more. The master type-founder usually supplies the printer with all the materials of his art, embracing not only types, leads, brass rules, and ordinary ornaments, but also cases, composing-sticks, galleys, printing-presses, and other articles too numerous to be mentioned.

7. The inventor of the art of casting types was Peter Shœffer, first servant or workman employed by Guttemburg and Faust. He privately cut a matrix for each letter of the alphabet, and cast a quantity of the types. Having shown the products of his ingenuity to Faust, the latter was so much delighted with the contrivance, that he made him a partner in the printing business, and gave him his only daughter, Christina, in marriage.

8. The character first employed was a rude old Gothic, mixed with secretary, designed on purpose to imitate the hand-writing of those times, and the first used in England were of this kind. To these succeeded what is termed *old English,* or *black letter,* which is still occasionally applied to some purposes; but Roman letter is now the national character not only of England, but of France, Spain, Portugal, and Italy. In Germany, and in the states surrounding the Baltic, letters are used which owe their foundation to the Gothic, although works are occasionally printed for the learned in Roman

9. The Roman letter owes it origin to the nation whence it derives its name, although the faces of the present and ancient Roman letters differ materially, on account of the improvements which they have undergone at various times. For the invention of the Italic character, we are indebted to Aldus Manutius, who set up a printing-office in Venice, in 1496, where he also introduced Roman types of a neater cut.

10. Before the American revolution, type-founding was carried on at Germantown, Pennsylvania, by Christopher Sower, at Boston by Mr. Michelson, and in Connecticut by Mr. Buel; but there was too little demand for types, to afford these enterprising individuals much patronage. Soon after the close of the revolution, John Baine established a foundery in Philadelphia. The printers, however, were not supplied with every necessary material and implement of the art from American founderies, until 1796, when Messrs. Binny & Ronaldson commenced the business in the same city. Baine and Ronaldson were both from Edinburgh, Scotland. The first type-foundery was established in New-York, in 1809, by Robert Lothian, a Scotch clergyman, and father of the ingenious type-founder, George B. Lothian.

11. In the year 1827, William M. Johnson, of New-York, invented the machine for casting types now used by John T. White, and in 1838, David Bruce, Junr., produced another, which was purchased by George Bruce. George B. Lothian has also lately invented a machine for the same purpose, and likewise one for reducing types to an equal thickness. Both of these machines act with great accuracy. There are now in the United States sixteen type-founderies; viz., two in Boston, six in New-York, three in Philadelphia, one in Baltimore, one in Pittsburg, one in Cincinnati, one in Louisville, and one in St. Louis.

THE STEREOTYPER.

1. The word *stereotype* is derived from two Greek words—*stereos*, solid, and *tupos*, a type. It is applied to pages of types in a single piece, which have been cast in moulds formed on common printing types or wood-cuts. They are composed of lead and antimony, in the proportion of about six parts of the former to one of the latter. Sometimes a little tin is added.

2. The types are *set up* by *compositors*, as usual in printing, and *imposed*, or locked up, one or several pages together, in an iron *chase* of a suitable size. Having been sent to the *casting-room*, the types are slightly oiled, and surrounded with a frame of brass or type-metal. They are then covered with a thin mixture of finely pulverized plaster and water. In about ten minutes, the plaster becomes hard enough to be removed.

3. The mould, thus formed, having been baked in an oven, is placed in an iron pan of an oblong shape, and sunk into a kettle of the melted composition above mentioned, which is admitted at the four corners of the cover to the cavities of the mould beneath. The pan is then raised from the kettle, and placed over water. When the metal has become cool, the contents of the pan are removed, and the plaster is broken and washed from the plate.

4. As fast as the pages are cast, they are sent to the *finishing-room*. Here they are first planed on the back with a machine, for the purpose of making them level and of an equal thickness. The letters are then examined, and, when deficient, are rendered perfect by little steel instruments called *picks*. Corrections and alterations are made by cutting out original lines, and inserting common printing types, or lines stereotyped for the purpose. The types are cut off close to the back with pincers, and fastened to the place with solder. The plates, when they are finished, are about one-sixth of an inch in thickness.

5. When all the pages of a work have been completed, they are packed in boxes, which are marked with certain letters of the alphabet, to indicate the form or pages which they contain. While the pages are applied in printing, they are fastened to blocks of solid wood, which, with the plates, are intended to be the same in height with common types.

6. The first stereotype plates were cast by J. Van der Mey, a Dutchman, who resided at Leyden about the year 1700. A quarto and folio Bible, and two or three small works, were printed from pages of his casting; but at his death, the art appears to have been lost, although the plates of these two Bibles are still extant, the former at Leyden, and the latter at Amsterdam.

7. In 1725, William Ged, of Edinburgh, without

knowing what had been done in Holland by Van der Mey, began to make stereotype plates. But being unable to prosecute the business alone for want of funds, he united in partnership with three others. One of the partners being a type-founder, supposing that success in the enterprise would injure his business, employed men to compose and print the proposed works in a manner that he thought most likely to spoil them.

8. Accordingly, the compositors, while correcting one error in the proof, made intentionally several more; and the pressmen battered the letter, while printing the books. By these dishonest and malicious proceedings, the useful enterprise of Mr. Ged was defeated. He, however, afterwards printed, in an accurate manner, two or three works. The first of these was a Sallust, the pages of which were set up by his son, James Ged, who was but an apprentice to the printing-business. This part of the work was performed in the night, when the workmen were absent from the office.

9. After the death of Mr. Ged, no attention was paid to the art, and a knowledge of it was lost at the decease of his son, which took place, about the year 1771: but it was a third time invented by Alexander Tilloch, Esq., who, in conjunction with Mr. Foulis, printer to the University of Glasgow, made many experiments, until plates were produced yielding impressions which could not be distinguished from those of the types from which they had been cast. But owing to circumstances unconnected with the real utility of the art, the business was not prosecuted to a great extent.

10. About the year 1804, the art was again revived by the late Earl Stanhope, assisted by Mr. A. Wilson, a printer, who turned his whole attention that way. In their efforts to complete the invention, they

were assisted by Messrs. Tilloch and Foulis; and, although they succeeded after many experiments, they were strenuously opposed in their efforts to introduce the practice, the printers supposing, perhaps with some reason, that it would prove injurious to their business.

11. This useful art was introduced into the United States by J. Watts, an Englishman from London, who had acquired a knowledge of the process from A. Wilson. He entered into a partnership with Joseph D. Fay and Pierre C. Van Wyck, Esquires. They first stereotyped the Westminster Catechism, which was printed by J. Watts & Co., for Messrs. Whiting & Watson, in 1813. They also stereotyped a New Testament. But the business proving to be unproductive, Fay and Van Wyck retired from the concern. Watts afterwards stereotyped about one third of an octavo Bible. The moulding of all the plates produced in Watts's foundery was executed by Mrs. Watts. On the 21st of March, 1815, Watts sold all his plates, together with his materials and knowledge of the process, to B. S. and J. B. Collins, for $6500. The Messrs. Collins afterwards carried on the business successfully.

12. In 1812, David Bruce went to England for the express purpose of obtaining a knowledge of the art, as it was kept a profound secret by Watts; and having learned the method of one Nicholson, of Liverpool, and having also acquired some knowledge of Earl Stanhope's plan, he returned to New-York, and commenced stereotyping, in conjunction with his brother, George Bruce, in the year 1813. They soon completed two setts of 12mo plates for the New Testament, one of which they sold to Matthew Carey, Nov. 8, 1814. Soon afterwards, they finished the whole Bible. David Bruce invented the machine for planing the plates, in 1815.

THE PAPER-MAKER, AND THE BOOKBINDER.

THE PAPER-MAKER.

1. The materials on which writing was executed, in the early days of the art, were the leaves and bark of trees and plants, stones, bricks, sheets of lead, copper, and brass, as well as plates of ivory, wooden tablets, and cotton and linen cloth.

2. The instruments with which writing was practised were adapted to the substance on which it was to be formed. The *stylus*, which the Romans employed in writing on metallic tablets covered with wax, was made of iron, acute at one end, for forming the letters, and flat or round at the other, for erasing what may have been erroneously written.

3. For writing with ink, the *calamus*, a kind of reed, sharpened at the point, and split like our pens

was used. Some of the Eastern nations still write with bamboos and canes. The Chinese inscribe their characters with small brushes similar to camel's hair pencils. We have no certain evidence of the application of *quills* to this purpose until the seventh century.

4. As the literature of antiquity advanced, a material adapted to works of magnitude became necessary, and this was found both in the skins of animals, and in the celebrated plant papyrus, of Egypt; but the time when they were first applied to this purpose cannot be determined, although it is probable that the former has the preference as regards priority.

5. The papyrus was an aquatic plant, which grew upon the banks of the Nile. In the manufacture of paper from this reed, it was divested of its outer covering, and the internal layers, or laminæ, were separated with the point of a needle or knife. These layers were spread parallel to each other on a table, in sufficient numbers to form a sheet; a second layer was then laid with the strips crossing those of the first at right angles; and the whole having been moistened with water, was subjected to pressure between metallic surfaces. The pressure, aided by a glutinous substance in the plant, caused the several pieces to become one uniform sheet.

6. Parchment was manufactured from the skins of sheep and goats. In the preparation, these were first steeped in water impregnated with lime, and afterwards stretched upon frames, and reduced by scraping with sharp instruments. They were finished by the application of chalk, and by rubbing them with pumice-stone. The skins of very young calves, dressed in a similar manner, was called vellum. Parchment and vellum are still used for deeds and other important documents.

7. When Attalus, about 200 years before Christ,

was about to found a library at Pergamus, which should rival that of Alexandria, one of the Ptolemies, then king of Egypt, jealous of his success, prohibited the exportation of papyrus; but the spirited inhabitants of Pergamus manufactured parchment as a substitute, and formed their library principally of manuscripts on this material. From this fact, it received the name of *Pergamena* among the Romans, who gave it also the appellation of *Membrana*.

8. The greatest quantity of paper was manufactured at Alexandria, and the commerce in this article greatly increased the wealth of that city. In the fifth century, paper was rendered very dear by taxation; and this probably was an inducement for an effort to produce a substitute. Accordingly, in the eighth century, it began to be superseded by cotton paper, although it continued in use in some parts of Europe, until three hundred years after the period last mentioned.

9. The manufacture of cotton paper was introduced into Spain, in the eleventh century, by the Arabians, who became acquainted with it in Bucharia as early as A.D. 704. About the year 1300, it was commenced in Italy, France, and Germany; and, in some of the paper-mills of these countries, paper was made from cotton rags. Linen paper is thought to have originated in Germany, about the year 1318.

10. The first paper-mill in England was erected by a German, named Spillman, in 1588; but no paper, except the coarse brown sorts, was made in that country, until about the year 1690. The finer kinds, both for writing and printing, were, before that time, imported from the Continent. But the paper of English manufacture will now compare with that of any other country. The French also make very fine paper.

11. In the United States, this manufacture has rapidly increased in amount within a few years. Ac-

cording to an estimate made in 1829, it appears that the whole annual product of the mills is worth between five and seven millions of dollars, and that the rags collected in this country amount to about two millions. The number of hands employed in the business are ten or eleven thousand, of whom about one-half were females. The manufacture has since been considerably increased, although the number of operatives may have been diminished, on account of the introduction of improved machinery.

12. Nature has supplied us with a great variety of substances from which paper may be fabricated, as flax, hemp, cotton, straw, grass, and the bark of several kinds of trees; but the fibres of the three first productions, in the form of rags, are the most usual materials. Most of these are primarily purchased from the people at large, by retail booksellers, country merchants, and pedlers, who in turn dispose of them to persons called rag-merchants, or directly to the paper-makers. When the rags come from the original collectors, all kinds are mixed together; but they are assorted according to their color and the nature of their original fibre, either by the rag-merchants, or by the paper-makers themselves.

13. In our attempts to afford the reader an idea of this manufacture in general, letter-paper has been selected, as affording the best means of illustration; since for this kind of paper, the best stock is employed, and the greatest skill is exerted in every stage of the process.

14. The process of the manufacture is commenced by cutting the rags into small pieces, by the aid of a sharp instrument, commonly a piece of a scythe, which is placed in a position nearly perpendicular before the operator. In the reduction of very coarse rags, such as sail-cloth, a cutting machine is also employed. Then, with the view of sifting out the loose

particles of dirt, the rags are deposited in a large octagonal sieve made of coarse wire, and placed in a close box in a horizontal position. The sieve is moved by machinery, like the bolt of a flour-mill.

15. The second stage of the process consists chiefly in the reduction of the rags to a *pulp*. This is effected by the action of a cutting machine, the essential parts of which are two sets of blunt knives, the one stationary, and the other revolving. The machine is placed in a large elliptical tub, in which the rags are also deposited, with a suitable quantity of water. The liquid and fibrous contents of the tub are kept moving in a circle by the action of the machine, through which it passes at one point of its revolution.

16. The maceration occupies from ten to twenty hours, according as the material is more or less rigid; and, during part of this time, water is permitted to run in at one side of the tub, and out at the other, to render the pulp perfectly clean. Towards the close of this process, the pulp, if necessary, is bleached by means of chloride of lime, and oil of vitriol. It is also sometimes colored by adding a quantity of dye-stuff. The bleaching and coloring are effected without interrupting the action of the machine. The rags having been thus reduced, the pulp, together with a suitable quantity of water, is let out into a reservoir, from which it is drawn off into a *vat*, as fast as it may be needed for the production of the paper.

17. With this vat is connected the paper-making machine; and the part of the latter which first comes in contact with the material is a hollow cylinder, surrounded with a fine web of wire-cloth. This cylinder being immersed in the contents of the vat more than one-half of its diameter. the water passes off with a uniform rapidity, and the fibrous particles which had been suspended in it, settle with a remarkable uniformity on the outside of the brazen web. As

the cylinder revolves, a continued sheet is produced, which is taken up by an endless web of woollen cloth, and carried round another cylinder of equal diameter, and then between two more, by which it is partially pressed.

18. From between these rollers, the paper passes out, in a continued sheet, upon a large cylindrical reel, called the *lay-boy;* and when a certain quantity of it, which is determined by a gauge, has been accumulated, the lay-boy is removed to a low table. The paper is then cut, with a toothless handsaw, into sheets twice the size of letter-paper. This part of the operation is very quickly performed, as a workman can cut up and pile in heaps, to be pressed, twenty reams in half that number of minutes, and attend to the machine at the same time.

19. After the paper has been successively pressed, and handled by separating the sheets two or three times, it is hung up on small poles, in an airy room, to be dried; and having been again pressed, it is sized by holding a quantity of the sheets at a time in a thin solution of glue and alum, the former of which is prepared in the paper-mill for the purpose, from shreds and parings of raw hides. The paper is freed from superfluous portions of the size, by submitting it to the action of a press. It is again dried as before, and again pressed; after which, the several sheets are examined, and freed from lumps and other extraneous substances.

20. They are then severed in half with a cutting machine, and afterwards calendered, by passing the sheets successively between rollers; or they are pressed between smooth pasteboards. In the latter case, hot metallic plates are sometimes interposed between every few quires of the sheets. The paper, when treated in this way, is called hot-pressed. It is next counted off into half-quires, put up into reams

pressed, trimmed, and finally enveloped in two thick sheets of paper, which completes the whole process of the manufacture.

21. The manufacture of paper, as just described, seems to be a tedious process; yet with two machines and a suitable number of hands, say sixty or eighty, three hundred reams of letter-paper can be produced from the raw material in a single day. It is hardly necessary to remark, that paper is of various qualities, from the finest bank-note paper, down to the coarsest kinds employed in wrapping up merchandise, and that, for every quality, suitable materials are chosen. The process of the manufacture is varied, of course, to suit the materials. None but writing and drawing paper requires to be sized.

22. Until after the beginning of the present century, paper was made exclusively *by hand*, and this method is still continued in a majority of the mills in the United States, although it is rapidly going out of use. It differs from that just described chiefly in the manner of collecting the pulp to form the paper, this being effected by means of a *mould*, a frame of wood with a fine wire bottom, of the size of the proposed sheet. In the use of this instrument, a quantity of the pulp is taken up, and while the *vatman*, or *dipper*, holds it in a horizontal position, and gives it a gentle shaking, the water runs out through the interstices of the wire, and leaves the fibrous particles upon the mould in the form of a sheet. The sheets thus produced are pressed between felts, and afterwards treated as if they had been formed by means of a machine.

23. The first idea of forming paper in a continued sheet originated in France; but a machine for this purpose is said to have been first made completely successful in England, by Henry and Sealy Fourdrinier. Many machines made after their model, as well as those of a different construction, are in use in the

United States, to some of which is attached an apparatus for drying, sizing, and pressing the paper, as well as for cutting it to the proper size. Very few machines, however, yield paper equal in firmness and tenacity to that produced by hand.

THE BOOKBINDER.

1. BOOKBINDING is the art of arranging the pages of a book in proper order, and confining them there by means of thread, glue, paste, pasteboard, and leather.

2. This art is probably as ancient as that of writing books; for, whatever may have been the substance on which the work was executed, some method of uniting the parts was absolutely necessary. The earliest method with which we are acquainted, is that of gluing the sheets together, and rolling them upon small cylinders. This mode is still practised in some countries. It is also everywhere used by the Jews, so far as relates to one copy of their law deposited in each of their synagogues.

3. The name Egyptian is applied to this kind of binding, and this would seem to indicate the place of its origin. Each volume had two rollers, so that the continued sheet could be wound from one to the other at pleasure. The square, or present form of binding, is also of great antiquity, as it is supposed to have been invented at Pergamus, about 200 years before Christ, by King Attalus, who, with his son Eumenes, established the famous library in that city.

4. The first process of binding books consists in folding the sheets according to the paging. This is done by the aid of an ivory knife, called a *folder;* and the operator is guided in the correct performance of the work by certain letters called *signatures,* placed at the bottom of the page, at regular intervals through the book.

5. Piles of the folded sheets are then placed on a long table in the order of their signatures, and gathered, one from each pile, for every book. They are next beaten on a stone, or passed between steel rollers, to render them smooth and solid. The latter method has been introduced within a few years. This operation certainly increases the intrinsic value of the book; but it is not employed in every case, since it is attended with some additional expense, and since it diminishes the thickness of the book, and consequently its value in the estimation of the public at large.

6. The sheets, having been properly pressed, are next sewed together upon little cords, which, in this application, are called *bands*. During the operation these are stretched in a perpendicular direction, at suitable distances from each other, as exhibited in the foregoing cut. The folded sheets are usually notched on the back by means of a saw, and at these points they are brought in juxta-position with the bands. After the pages of several volumes have been accumulated, the bands are severed between each book. The folding, gathering, and sewing, are usually performed by females.

7. At this stage of the process, the books are received by the men or boys, who generally *take on* one hundred at a time. The workman first spreads some glue on the backs of each book with a brush. He then places them, one after the other, between boards of solid wood, and beats them on the back with a hammer. By this means the back is rounded, and a groove formed on each side for the admission of one edge of the pasteboards.

8. These having been applied, and partially fastened by means of the bands, which had been left long for the purpose, the books are pressed, and the leaves of which they are composed are trimmed with an instrument called a *plough*. The pasteboards are also cut

to the proper size by the same means, or with a huge pair of shears. In the preceding picture, a workman is represented at work with the plough. The edges are next sprinkled with some kind of coloring matter, or covered with gold leaf. A strip of paper is then glued on the back, and a *head-band* put upon each end.

9. The book is now ready to be covered. This is done either with calf, sheep, or goat skin, or some kind of paper or muslin; but, whatever the material may be, it is cut into pieces to suit the size of the book; and, having been smeared on one side with paste, if paper or leather, or with glue, if muslin, it is drawn over the outsides of the pasteboards, and doubled in upon the inside.

10. The covers, if calf or sheep skin, are next sprinkled or marbled. The first operation is performed by dipping the brush in a kind of dye, made for the purpose, and beating it with one hand over a stick held in the other; the second is performed in the same manner, with the difference that they are sprinkled first with water, and then with the coloring matter.

11. After a small piece of morocco has been pasted on the back, on which the title is to be printed in gold leaf, and one of the waste leaves has been pasted down on the inside of each of the covers, the books are pressed for the last time. They are then glazed by applying the white of an egg with a sponge.

12. The books are now ready for the reception of the ornaments, which consist chiefly of letters and other figures in gold leaf. In executing this part of the process, the workman cuts the gold into suitable strips or squares on a cushion.

13. These are laid upon the books by means of a piece of raw cotton, and afterwards impressed with types moderately heated over a charcoal fire; or the

strips of gold are taken up, and laid upon the proper place with instruments called *stamps* and *rolls*, which have on them figures in relief. The portion of the leaf not impressed with the figures on the tools, is easily removed with a silk rag. The books are finished by applying to the covers the white of an egg, and rubbing them with a heated steel *polisher*.

14. The process of binding books, as just described, is varied, of course, in some particulars, to suit the different kinds of binding and finish. A book stitched together like a common almanac, is called a pamphlet. Those which are covered on the back and sides with leather, are said to be *full-bound;* and those which have their backs covered with leather, and the sides with paper, *half-bound.*

15. The different sizes of books are expressed by terms indicative of the number of pages printed on one side of a sheet of paper; thus, when two pages are printed on one side, the book is termed a folio; four pages, a quarto; eight pages, an octavo; twelve pages, a duodecimo; eighteen pages, an octodecimo. All of these terms, except the first, are abridged by prefixing a figure or figures to the last syllable: thus, 4to for quarto, 8vo for octavo, 12mo for duodecimo, &c.

16. The manufacture of account-books, and other blank or *stationary* work, constitutes an extensive branch of the bookbinder's business. It is not necessary, however, to be particular in noticing it, as the general process is similar to that of common bookbinding. Those binders who devote much attention to this branch of the trade, have a machine by which paper is ruled to suit any method of keeping books, or any other pattern which may be desired.

THE BOOKSELLER.

1. THE book-trade has arisen from small beginnings to its present magnitude and importance. Before the invention of typography, it was carried on by the aid of transcribers; and the booksellers of Greece, Rome, and Alexandria, during the flourishing state of their literature, kept a large number of manuscript copyists in constant employ. Among the Romans, the transcribers or copyists were chiefly slaves, who were very valuable to their owners, on account of their capacity for this employment.

2. In the middle ages, when learning was chiefly confined to the precincts of monastic institutions, the monks employed much of their time in copying the ancient classics and other works; and this labor was often imposed upon them as a penance for the com-

mission of sin. From this cause, and from an ignorance of the true meaning of the author, much of their copying was inaccurately performed, so that great pains have been since required in the correction of the manuscripts of those times.

3. This mode of multiplying copies of books was exceedingly slow, and, withal, so very expensive, that learning was confined almost exclusively to people of rank, and the lower orders were only rescued from total ignorance by the reflected light of their superiors. For a long time, during the reign of comparative barbarism in Europe, books were so scarce, that a present of a single copy to a religious house was thought to be so valuable a gift, that it entitled the donor to the prayers of the community, which were considered efficacious in procuring for him eternal salvation.

4. After the establishment of the universities of Paris and Bologna, there were dealers in books, called *stationarii*, who loaned single manuscripts at high prices; and, in the former place, no person, after the year 1432, could deal in books in any way, without permission from the university, by which officers were appointed to examine the manuscripts, and fix the price for which they might be sold or hired out.

5. For a long time after the invention of printing, the printers sold their own publications; and, in doing this, especially at some distance from their establishments, they were aided by those who had formerly been employed as copyists. Some of these travelling agents, at length, became stationary, and procured the publication of works on their own account.

6. The first bookseller who purchased manuscripts from the authors, and caused them to be printed without owning a press himself, was John Otto, of Nuremburg. He commenced this mode of doing business, in 1516. In 1545, there were, for the first time, two such booksellers in Leipsic. The great mart for the

sale of their books was Frankfort on the Maine, where were held three extensive fairs every year. Leipsic, however, soon became, and still continues, the centre of the German book-trade.

7. The first Leipsic catalogue of books appeared as early as the year 1600; but the fairs at that place did not become important, as regards the book-trade, until 1667, when it was attended by nineteen foreign booksellers. The booksellers of Germany, as well as some from distant countries, meet at the semi-annual fairs held in that city, to dispose of books, and to settle their accounts with each other. Every German publisher has also an agent there, who receives his publications, and sends them, according as they are ordered, to any part of Germany.

8. In no other part of the world, has such a connexion of booksellers been formed, although almost every kingdom of Europe has some city or cities in which this branch of trade is chiefly concentrated; as London, in England; Edinburgh, in Scotland; and Amsterdam, Utrecht, Leyden, and Haerlem, in the Netherlands. In Spain and Portugal, the price of every book is regulated by the government.

9. A very convenient method of effecting the sale and exchange of books among booksellers, has been adopted in the United States; and this is by auction. A sale of this kind is held in Boston once, and in New-York and Philadelphia twice, every year; and none are invited to attend it but the *trade;* hence such sales are denominated *trade-sales.*

10. The sale is usually conducted by an auctioneer who has been selected by a committee of the trade in the city in which it is to be held. In order to obtain a sufficient amount of stock for the purpose, the agent issues proposals, in which he informs publishers and others concerned in this branch of business, of his intention, and solicits invoices of books, to be sold at

the time specified. A catalogue of all the books thus sent for sale, is distributed among the booksellers.

11. The booksellers having assembled, the books which may have been accumulated from different parts of the Union, are offered in convenient lots, and *struck off* to the highest bidder. Each purchaser holds in his hand the printed catalogue, on the broad margin of which he marks, if he sees fit, the prices at which the books have been sold; and the record thus kept affords a tolerable means of determining their value, for a considerable time afterwards.

12. A sale of this kind occupies from four to six days; and, at the close of it, a settlement takes place, in which the parties are governed by the terms previously published. The payments are made in cash, or by notes at four or six months, according to the amount which the purchaser may have bought out of one invoice. The conductors of the sale are allowed about five per cent. commission for their services.

13. A vast number of books is also sold, every year, at auction, to miscellaneous collections of people, not only in the cities and considerable towns, but likewise in the villages throughout the country. By many booksellers, this method of sale is thought to be injurious to the trade, since it has reduced the prices of books, and interfered with the regular method of doing business. These disadvantages, however, have been far overbalanced by the increased number of readers which has been thus created.

14. The circulation of books is likewise promoted by means of travelling agents, who either sell them at once, or obtain subscriptions for them with the view to their future delivery. These methods have been employed more or less from the very commencement of the printing business; and they have probably contributed more to the general extension of knowledge than the sale of books by stationary booksellers. In

fact, they are among the most prominent causes of the vast trade in books, which is now carried on, especially in the United States.

15. Nevertheless, publishers, who do not employ agents to vend their books, generally consider them interlopers upon their business; and the people themselves, who owe a great share of their intellectual cultivation to this useful class of men, are generally averse to afford them the necessary patronage, because they require a small advance on the city prices to pay travelling expenses.

16. A considerable amount of books is also sold by merchants who reside at some distance from the cities and large towns. They, however, seldom venture to purchase those which have not been well known and approved in their neighborhood; and, in a majority of cases, regard them as mere subjects of merchandise, without taking into consideration the effects most likely to be produced by these silent, but powerful agents, when circulated among their customers.

17. Some booksellers in Europe confine their trade chiefly to particular departments; such as law, theology, and medicine. Others deal in toy-books, and books of education, or in rare and scarce books. This is the case, to a limited extent, in the United States, although our booksellers commonly keep an assortment of miscellaneous publications, as well as various articles in the stationary line; such as paper, quills, inkstands, and blank work.

THE ARCHITECT.

1. ARCHITECTURE, in the general sense of the word, is the art of planning and erecting buildings of all kinds, whether of a public or private nature; and it embraces within its operations a variety of employments, at the head of which must be placed the Architect. Architecture is of several kinds, such as *civil, naval, military,* and *aquatic;* but it is the first only that we propose to notice in the present article.

2. The construction of buildings as means of shelter from the weather, appears to have been among the earliest inventions; and, from the skill exhibited in the construction of the ark, we have reason to believe that architecture had been brought to considerable perfection before the deluge. This opinion is also supported by the fact stated in holy writ, that the

descendants of Noah, not more than one hundred years after the great catastrophe just mentioned, attempted to build a city and a lofty tower with bricks burned in the fire. This project could never have been thought of, had they not been influenced by the knowledge of former centuries.

3. The confusion of the language of the people caused their dispersion into different parts of the earth; and, in their several locations, they adopted that method of constructing their dwellings, which the climate required, and the materials at hand admitted; but, whatever the primitive structure may have been, it was continued, in its general features, from age to age, by the more refined and opulent inhabitants; hence the different styles of building, which have been continued, with various modifications, to the present day.

4. The essential elementary parts of a building are those which contribute to its support, inclosure, and covering; and of these the most important are the foundation, the column, the wall, the lintel, the arch, the vault, the dome, and the roof. Ornamental and refined architecture is one of the fine arts; nevertheless, every part of an edifice must appear to have utility for its object, and show the purpose for which it has been designed.

5. The *foundation* is usually a stone wall, on which the superstructure of the building rests. The most solid basis on which it is placed is rock, or gravel which has never been disturbed; next to these are clay and sand. In loose or muddy situations, it is always unsafe to build, unless a solid basis can be artificially produced. This is often done by means of timber placed in a horizontal position, or by driving wooden piles perpendicularly into the earth; on a foundation of the latter description, the greater part of the city of Amsterdam has been built.

6. The *column*, or *pillar*, is the simplest member of a building, although it is not essential to all. It is not employed for the purpose of inclosure, but as a support to some part of the superstructure, and the principal force which it has to resist is that of perpendicular pressure. The column is more frequently employed in public than in private buildings.

7. The *wall* may be considered the lateral continuation of the column, answering the purposes both of support and inclosure. It is constructed of various materials, but chiefly of brick, stone, and marble, with a suitable proportion of mortar or cement. Walls are also made of wood, by first erecting a frame of timber and then covering it with boards; but these are more perishable materials, which require to be defended from the decomposing influence of the atmosphere, by paint or some other substance.

8. The *lintel* is a beam extending in a right line from one column or wall to another over a vacant space. The *floor* is a lateral continuation or connexion of beams, by means of a covering of planks. The strength of the lintel, and, in fact, every other elementary part of a building used as a support, can be mathematically determined by the skilful architect.

9. The *arch* answers the same purpose as the lintel, although it far exceeds it in strength. It is composed of several pieces of a wedge-like form, and the joints formed by the contact of flat surfaces point to a common centre. While the workmen are constructing the arch, the materials are supported by a *centring* of the shape of its internal surface. The upper stone of an arch is called the *key-stone*. The supports of an arch are called *abutments*; and a continuation of arches, an *arcade*.

10. The *vault* is the lateral continuation of an arch, and bears the same relation to it that a wall bears to a column. The construction of a simple

vault is the same with that of an arch, and it distributes its pressure equally along the walls or abutments. A complex or groined vault is made by the intersection of two of the common kind. The groined vault is much used in Gothic architecture.

11. The *dome*, or *cupola*, is a hemispherical or convex covering to a building or a part of it. When built of stone it is a very strong kind of structure, even more so than the arch, since the tendency of the parts to fall is counteracted by those above and below, as well as by those on each side. During the erection of the cupola, no centring is required, as in the case of the arch.

12. The *roof* is the most common and cheap covering to buildings. It is sometimes flat, but most commonly oblique, in shape. A roof consisting of two oblique sides meeting at the top, is denominated a *pent* roof; that with four oblique sides, a *hipped* roof; and that with two sides, having each two inclinations of different obliquities, a *curb* or *mansard* roof. In modern times, roofs are constructed of wood, or of wood covered with some incombustible material, such as tiles, slate, and sheets of lead, tin, or copper. The elementary parts of buildings, as just described, are more or less applicable in almost every kind of architecture.

13. The architecture of different countries has been characterized by peculiarities of form and construction, which, among ancient nations, were so distinct, that their edifices may be identified at the present day even in a state of ruin; and, although nearly all the buildings of antiquity are in a dilapidated state, many of them have been restored, in drawings and models, by the aid of the fragments which remain.

14. The different styles of building which have been recognised by the architect of modern times, are, the Egyptian, the Chinese, the Grecian, the Ro-

man, the Greco-Gothic, the Saracenic, and the Gothic. In all these, the pillar, with its accompaniments, makes a distinguished figure. The following picture has therefore been introduced by way of explanation. The columns are of the Corinthian order of architecture.

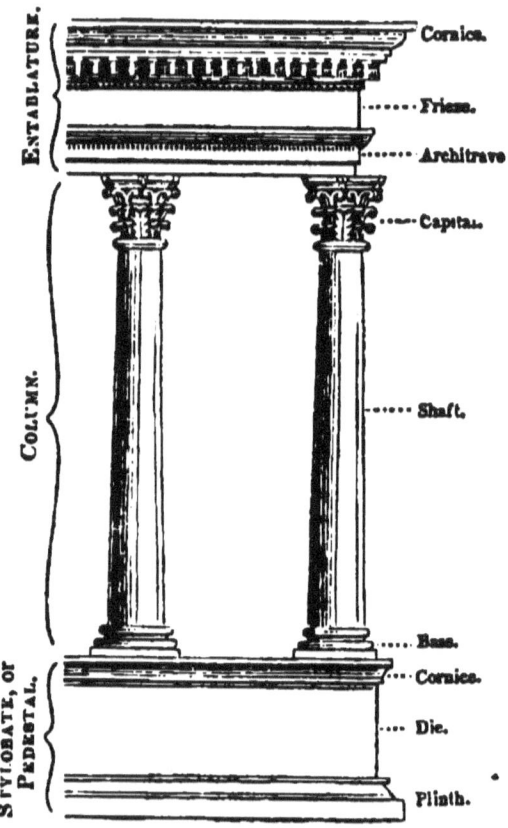

15. The *Egyptian style.*—The first inhabitants of Egypt lived in mounds, caverns, and houses of mud; and, from these primitive structures, the Egyptians, at a later period, derived their style of architecture. The walls of their buildings were very thick, and sloping on the outside; the roof was flat, and composed of blocks of stone, extending from one wall or pillar to another; and the columns were short and

large, being sometimes ten or twelve feet in diameter. Pyramids of prodigious magnitude, and obelisks composed of a single stone, sometimes often exceeding seventy feet in height, are structures peculiarly Egyptian. The architecture of the Hindoos seems to have

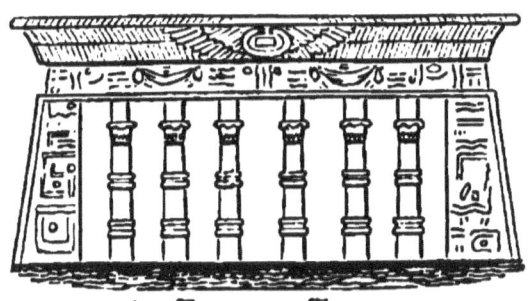

AN EGYPTIAN TEMPLE.

been derived from primitive structures of a similar character.

16. *The Chinese style.*—The ancient Tartars, and other wandering tribes of Asia, appear to have lived in tents; and the Chinese buildings, even at the present day, bear a strong resemblance to these original

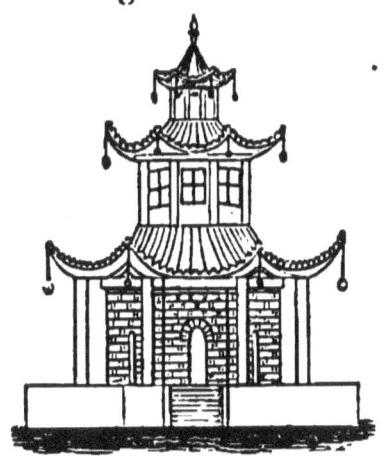

A CHINESE PAGODA.

habitations, since their roofs are concave on the upper side, as if made of canvas instead of wood.

Their porticoes resemble the awnings spread out on our shop-windows in the summer. The Chinese build chiefly of wood, although they sometimes use brick and stone.

17. *The Grecian style.*—This style of building had its origin in the wooden hut or cabin, the frame of which primarily consisted of perpendicular posts, transverse beams, and rafters. This structure was at length imitated in stone, and by degrees it was so modified and decorated in certain parts, as to give rise to the several distinctions called orders of architecture. The Greeks, in perfecting their system of architecture, were probably aided by Egyptian examples, although they finally surpassed all other nations in this important art.

18. *Orders of architecture.*—By the architectural orders are understood certain modes of proportioning and decorating the column and entablature. They were in use during the best days of Greece and Rome, for a period of six or seven centuries. The Greeks had three orders, called the *Doric*, the *Ionic*, and the *Corinthian*. These were adopted and modified by the Romans, who also added two others, called the *Tuscan* and the *Composite*.

19. *Doric order.*—The Doric is the oldest and most massive order of the Greeks. The column, in the examples at Athens, is about six of its diameters in height; in those of an earlier date, it is but four or five. The temple here adduced to illustrate this or-

THE TEMPLE OF THESEUS.

der was built by Cimon, son of Miltiades, about the

year 450 before Christ. It is said to be in a state of better preservation than any other of the ancient Greek edifices at Athens. It will be seen that the shafts are *fluted*, that is, cut in semicircular channels, in a longitudinal direction. The United States' Bank, at Philadelphia, is a noble specimen of this order.

20. *Ionic order.*—This order is lighter than the Doric, its column being eight or nine diameters in height. Its shaft has twenty-four or more flutings, separated from each other by square edges; and its capital consists, in part, of two double scrolls, called *volutes*, usually occupying opposite sides. These volutes are supposed to have been copied from ringlets of hair, or from the horns of the god Jupiter Ammon. The following example of this order consists of three temples, each of which was dedicated to a different individual, viz., Erectheus, Minerva Polias, and the nymph Pandrosus.

THE ERECTHEUM AT ATHENS.

21. *Corinthian order.*—The Corinthian is the lightest and most decorated of all the Grecian orders. Its column is usually ten diameters in height, and its shaft is fluted like that of the Ionic. Its capital is shaped like an inverted bell, and was covered on the outside with two rows of the leaves of the plant acanthus, above which are eight pairs of small volutes. It is said that this beautiful capital was suggested to the sculptor Callimachus by the growth of an acanthus about a basket, which had been accidentally left in a garden.

22. The Greeks sometimes departed so far from the strict use of their orders, as to employ the statues of slaves, heroes, and gods, in the place of columns. A specimen of this practice is exhibited in the cut illustrative of the Ionic order. It belongs to the temple dedicated to Pandrosus.

23. The most remarkable buildings of the Greeks were their temples. The body of these edifices consisted of a walled cell, usually surrounded by one or more rows of pillars. Sometimes they had a colonnade at one end only, and sometimes at both ends. Their form was generally oblong, and as the cells were intended as places of resort for the priests rather than for assemblies of the people, they were but imperfectly lighted. Windows were seldom employed; and light was admitted at the door at one end, or through an opening in the roof.

24. Grecian architecture is supposed to have been at its greatest perfection in the days of Pericles and Phidias, when sculpture is admitted to have attained its highest excellence. It was distinguished, in general, by simplicity of structure, fewness of parts, absence of arches, and lowness of pediments and roofs.

25. *Roman style.*—The Romans adopted the three Grecian orders, with some modifications; and also added two others, called the Tuscan and Composite. The former of these they borrowed from the nation whose name it bears, and the latter they formed by uniting the embellishments of the Doric and the Corinthian. The favorite order in Rome and its colonies was the Corinthian. Examples of single pillars of these orders may be seen at the end of this article.

26. The temples of the Romans generally bore a strong resemblance to those of the Greeks, although they often differed from the specimens of that nation in several particulars. The stylobate of the latter was usually a succession of platforms, which likewise

II.—I

served the purposes of steps, by which the building was approached on all sides. Among the Romans, it was usually an elevated structure, like a continued pedestal, on three sides, and accessible in front by means of steps. The dome was also very commonly employed rather than the pent roof. The following is an example of a temple at Rome.

TEMPLE OF ANTONIUS AND FAUSTINA.

27. *Greco-Gothic style.*—After the dismemberment of the Roman empire, the practice of erecting new buildings from the fragments of old ones became prevalent. This gave rise to an irregular style of building, which continued in use during the dark ages. It consisted of Greek and Roman details combined under new forms, and piled up into structures wholly unlike the original buildings from which the materials had been taken. Hence the appellations *Greco-Gothic* and *Romanesque* have been applied to it. The effect of this style of building was very imposing, especially when columns and arches were piled upon each other to a great height.

28. *Saracenic style.*—This appellation has been given to the style of building practised by the Moors

and Saracens in Spain, Egypt, and Turkey. It is distinguished, among other things, by an elliptical form of the arch. A similar peculiarity exists in the domes of the Oriental mosques, which are sometimes large segments of a sphere, appearing as if inflated; and at other times, they are concavo-convex on the outside. Several of these domes are commonly placed upon one building. The *minaret* is a tall slender tower, peculiar to Turkish architecture.

29. *Gothic style.*—The Goths, who overran a great part of the Western empire, were not the inventors of the style of architecture which bears their name. The term was first applied with the view to stigmatize the edifices of the middle ages, in the construction of which, the purity of the antique models had not been regarded. The term was at first very extensive in its application; but it is now confined chiefly to the style of building which was introduced into various parts of Europe six or eight centuries ago, and which was used in the construction of cathedrals, churches, abbeys, and similar edifices.

GOTHIC CATHEDRAL AT YORK.

30. The Gothic style is peculiarly and strongly marked. Its principles seem to have originated in the imitation of groves and bowers, under which the Druid priests had been accustomed to perform their sacred rites. Its characteristics are, pointed arches,

pinnacles and spires, large buttresses, clustered pillars, vaulted roofs, and a general predominance of the perpendicular over the horizontal.

31. The ecclesiastical edifices of this style of building are commonly in form of a cross, having a tower, lantern, or spire, erected at the point of intersection. The part of the cross situated towards the west is called the *nave;* the eastern part, the *choir;* and the transverse portion, the *transept.* A glance at the following diagram will enable the reader to understand the form of the ground-work more fully.

32. Any high building erected above a roof is called a *steeple,* which is also distinguished by different appellations, according to its form: if it is square topped, it is a *tower;* if long and acute, a *spire;* or if short and light, a *lantern.* Towers of great height in proportion to their diameter are denominated *turrets.* The walls of Gothic churches are supported on the outside by lateral projections, called *buttresses,* which extend from the bottom to the top, at the corners and between the windows. On the top of these are slender pyramidal structures or spires, called *pinnacles.* The summit or upper edge of a wall, if straight, is called a *parapet;* if indented, a *battlement.*

33. Gothic pillars or columns are usually clustered, appearing as if a number were bound together. They are confined chiefly to the inside of buildings, and are generally employed in sustaining the vaults which support the roof. The parts which are thrown

out of a perpendicular to assist in forming these vaults, have received the appellation of *pendentives*. The Gothic style of building is more imposing than the Grecian; but architects of the present day find it difficult to accomplish what was achieved by the builders of the middle ages.

34. In the erection of edifices at the present day, the Grecian and Gothic styles are chiefly employed, to the exclusion of the others, especially in Europe and America. Modern dwelling-houses have necessarily a style of their own, so far as relates to stories, windows, and chimneys; and no more of the styles of former ages can be applied to them, than what relates to the unessential and decorative parts.

110 THE ARCHITECT.

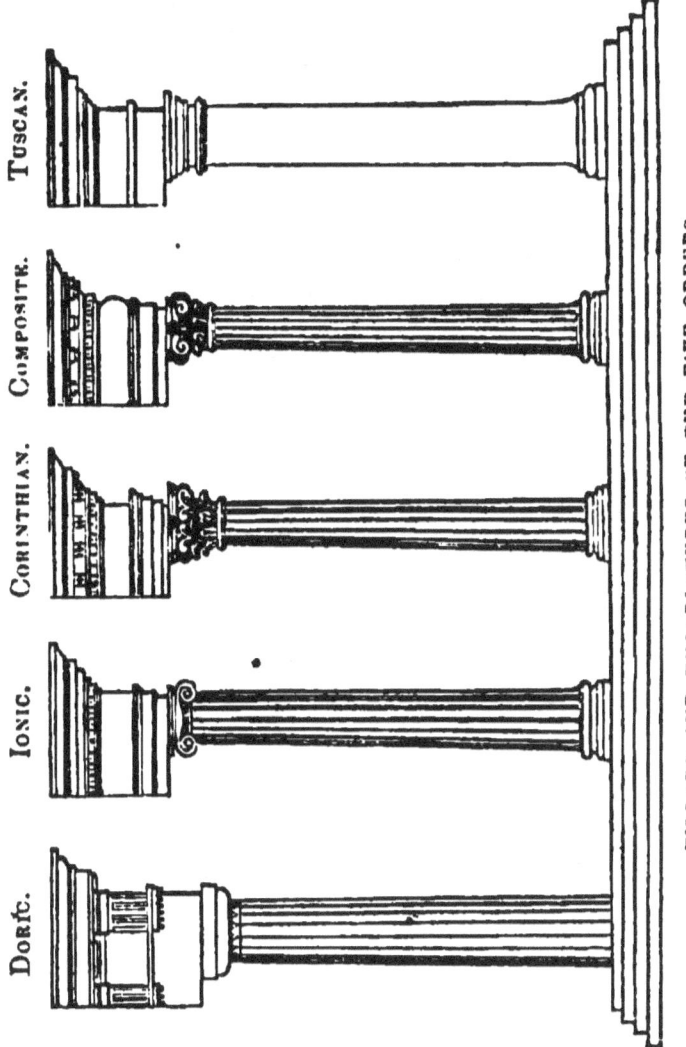

THE CARPENTER.

1. It is the business of the carpenter to cut out and frame large pieces of timber, and then to join them together, or fit them to brick or stone walls, to constitute them the outlines or skeleton of buildings or parts of buildings.

2. The joiner executes the more minute parts of the wood-work of edifices, comprehending, among other things, the floors, window-frames, sashes, doors, mantels, &c. Carpentry and joinery, however, are so nearly allied to each other, that they are commonly practised by the same individuals; and, in this article, they will be treated together.

3. Carpentry and joinery, as well as all other trades connected with building, are subservient to the architect, when an individual of this particular profes-

sion has been employed; but it most commonly happens, that the master-carpenter acts in this capacity. This is especially the case in the erection of common dwellings, and, in fact, of other edifices where nothing very splendid is to be attempted. It is to be regretted, however, that the professional architect has not been oftener employed; for, had this been the case, a purer taste in building would have generally prevailed.

4. Contracts for the erection of buildings are often made with the carpenter, as master-builder or architect. In such cases, it is his business to employ persons capable of executing every kind of work required on the proposed edifice, from the bricklayer and stone-mason to the painter and glazier. It not unfrequently happens, however, that the person himself, who proposes to erect a building, chooses to employ the workmen in the different branches.

5. The constituent parts of buildings having been explained in the article on architecture, it is unnecessary to enter here into minute details on this point; nor would a particular description of the various operations of the carpenter and joiner be useful to the general reader, since, in every place, means are at hand by which a general view of this business may be obtained by actual inspection.

6. The carpenter and joiner are guided, in the performance of their work, by well-defined rules, drawn chiefly from the science of Geometry, and which they have learned from imitation and practice, as well as, in many cases, from the valuable works which have been published on these branches of the art of building.

7. The principal tools with which they operate are the axe, the adze, the saw, the auger, the gauge, the square, the compasses, the hammer, the mallet, the crow, the rule, the level, the maul, and the plane; and of many of these there are several kinds.

8. The timbers most employed in building in the United States are chiefly pine, oak, beech, black walnut, cypress, larch, white cedar, and hemlock; but of these pine is in the greatest use. Oak and beech are much used in constructing frames, in which great strength is required. Of the pine, there are several species, of which the white and yellow are the most valuable; the former of these grows in the greatest abundance in the Northern, and the latter, in the Southern states.

9. Vast quantities of timber are annually cut into boards in saw-mills, and floated down the rivers from the interior, during the time of high water in the spring and fall, and sometimes at other seasons of the year. The boards, or, as they are frequently denominated, planks, are placed in the water, one tier above another, and fastened together with wooden pins. Several of such *rafts* are connected by means of withes to form one; and, at each end of this, are placed one or two huge oars, with which it may be guided down the stream. Upon these rafts, shingles and laths are also brought to market.

10. Logs and scantling to be employed in the frames of buildings are also conveyed down the rivers in the same manner. The business connected with the production of shingles, laths, boards or planks, and staves, is called lumbering; and it is carried on, more or less extensively, in the regions near the sources of all the large rivers in the United States, and in the British possessions in North America.

11. The trade in lumber has also given rise to another class of men, called lumber merchants; these purchase the lumber from the original proprietors, who bring it down the rivers, and, in their turn, sell it to builders and others. The lumbering business employs a large capital, and a numerous class of our citizens.

THE STONE-MASON, THE BRICKMAKER, &c.

THE MASON.

1. The art of Masonry includes the sawing and cutting of stones into the various shapes required in the multiplied purposes of building, and in placing them in a proper manner in the walls and other parts of edifices. It is divided into two branches, one of which consists in bringing the stones to the desired form and polish, and the other, in laying them in mortar or cement.

2. The rocks most used in building in the United States, are marble, granite, greenstone, sienite, soapstone, limestone, gypsum, and slate. These are found in a great many localities, not only on this continent, but on the other side of the Atlantic. Of these stones, there are many varieties, which are frequently designated by their sensible qualities, or by the name of

the place or country whence they are obtained; as *variegated, Italian, Egyptian,* or *Stockbridge marble,* and *Quincy stone.*

3. *The Stone-cutter.*—Stone-cutters procure their materials from the *quarry-men,* whose business it is to *get out* the stones from the quarries, in which they lie in beds, consisting either of strata piled upon each other, or of solid masses. Stones of any desirable dimensions are detached from the great mass of rock, by first drilling holes at suitable points, and then driving into them wedges with a sledge. These blocks are usually removed from the quarries, and placed on vehicles of transportation, by means of huge cranes, with which is connected suitable machinery.

4. The blocks of stone, received in their rough state by the stone-cutter, are divided, if required, into pieces of smaller size, by means of a toothless saw, aided by the attrition of sand and water. The other rough sides of the blocks are reduced to the proper form by means of steel *points* and *chisels* driven with a mallet. A kind of hammer with a point or chisel-like edge, is also used to effect the same object, especially in the softer kinds of stone.

5. For some purposes, the stones are required to be polished. This is especially the case with those employed in the ornamental parts of buildings. In the execution of this part of the work, the surface is rubbed successively with sand, freestone, pumice-stone, Scotch stone, crocus, and putty. When the face is a plane, the sand is applied by means of another stone, which is moved backwards and forwards upon it. In this way, two surfaces are affected at the same time.

6. In polishing irregular surfaces, the different kinds of stone are used in masses of convenient size; and the part applied to the surface to be polished is first brought to a form corresponding to it

The putty is an oxyde of tin, in form of powder. Crocus is the peroxyde of iron. The building-stone capable of receiving the highest polish is marble; and it is on this material that the stone-cutter, and the architectural carver or sculptor, exert their utmost skill; but some of the other stones which have been mentioned, possess the same quality to a considerable extent.

7. Carving architectural ornaments, such as pillars with their capitals, is a refined branch of this business; or it may rather be considered, of itself, a branch of sculpture. In the execution of this kind of work, the operator is guided by patterns, formed from the well-defined rules of the science of building. Very few stone-cutters attempt the execution of work so very difficult.

8. From the manufacture of mantel-pieces and monuments for the dead, the stone-cutter derives a great proportion of his profits. This will be manifest even to the superficial observer who may visit a few of the many stone-cutters' yards, to be found in any of our large cities. In some of these, blocks of marble are cut into slabs by the aid of steam-power.

9. In districts of country, also, where valuable stone is abundant, water is extensively employed for the same purpose. This is especially the case in Berkshire county, Massachusetts, where marble of a good quality is abundant. A great proportion of the marble slabs used by the stone-cutter are obtained from such mills. Some other operations of this business are also sometimes performed by the aid of machinery.

THE STONE-MASON.

1. In Philadelphia, and in many other cities not only in this country, but also in Europe, the stone-cutters *set their own work;* and this practice has led to

the habit of applying the term stone-mason to both stone-cutters and those who lay stone in mortar and cement. In New-York, however, as well as in some of the cities farther east, these two employments are kept more distinct. The stone-cutters in Philadelphia are sometimes denominated marble-masons.

2. But, in every city, there are persons called stone-masons, whose business consists exclusively in constructing the walls and some other parts of buildings with stone; and their operations are considerably enlarged in those places where there are no marble-masons. In many cases, the bricklayer is also so far a stone-mason, as to lay the foundation-walls of the buildings which he may erect. This is especially the case in the country, where the divisions of labor are not so minute as in cities. It may be well here to remark, also, that the bricklayers, in some places, perform the services of the marble-mason.

3. The marble-mason, in joining together several pieces in a monument, employs a kind of cement composed of about six parts of lime, one of pure sand, a little plaster, and as much water as may be necessary to form it to the proper consistency. No more of this cement is used than is required to hold the blocks or parts together, as one great object of the artist is to hide the joints as much as possible. The substance thus interposed, becomes as hard as the marble itself.

4. The cement employed in laying marble in common or large edifices, is somewhat different from that just described, as it consists of about three fourths of lime and one of sand. The latter substance is obtained, in an unmixed state, on the bays in every part of the world; hence it has received the appellation of *bay sand*.

5. When it cannot be conveniently had in a pure state, particles of the same kind can be separated in

sufficient quantities from their admixture with other substances. This is effected by sifting the compound through a sieve, into a small stream of water, which carries off the lighter particles that are unfit for use, whilst the sand, by its superior specific gravity, sinks to the bottom. The part which may be too coarse, remains in the sieve. This, however, except the rubbish, can be used in the coarser kinds of masonry.

6. The mortar, used in laying bricks and common stone, has a greater proportion of sand, which is generally of an inferior quality. Besides, the materials are incorporated with less care. Lime for the purposes of building is procured chiefly by calcining limestone in a kiln, with wood, coal, or some other combustible substance. It is also obtained by burning chalk, marble, and marine shells. Water poured upon newly-burnt or *quick* lime, causes it to swell, and fall to pieces into a fine powder. In this state it is said to be *slacked*.

7. Masonry is often required in situations under water, especially in the construction of bridges and locks of canals. Common mortar resists the action of the water very well, when it has become perfectly dry; yet, if it is immersed before it has had time to harden, it dissolves, and crumbles away.

8. The ancient Romans, who practised building in the water to a great extent, discovered a material, which, when incorporated with lime, either with or without sand, possessed the property of hardening in a few minutes even under water. This was a kind of earth found at Puteoli, to which was given the name of *pulvis puteolanus,* and which is the same now called *puzzolana.*

9. A substance denominated *tarras, terras,* or *tras,* found near Andernach, in the vicinity of the Rhine, possesses the same quality with puzzolana. It is this material which has been principally employed by the

Dutch, whose aquatic structures are superior to those of any other nation in Europe. Various other substances, such as baked clay and calcined greenstone, reduced to powder, afford a tolerable material for water-cements. Several quarries of water lime, which is similar in appearance to common limestone, has been lately discovered in the United States, which, being finely pulverized and mixed with sand, makes very good water-cement.

10. In buildings constructed with marble and other costly stones, the walls are not composed of these materials in their entire thickness; but, for the sake of cheapness, they are formed on the inside with bricks, commonly of a poor quality, so that in reality they can be considered only brick walls faced with stone. These two kinds of materials have no other connexion than what is produced by the mortar which may have been interposed, and the occasional use of clamps of iron. Such walls are said to be liable to become convex outwardly from the difference in the shrinking of the cement employed in laying the two walls.

11. The principal tools employed in cutting and laying stone are the saw, various kinds of steel points, chisels and hammers, the mallet, the square, the compasses, the level, the plumb-rule, the trowel, and the hod, to which may be added, the spade and the hoe. The last three instruments, however, are handled almost exclusively by laborers.

12. Besides these, contrivances are required to raise heavy materials to the various positions which they are to occupy. These consist, for the most part, of one or two shafts, commonly the mast of an old vessel, to which are attached tackle extending in various directions, and also those by which the blocks are to be raised. The rope belonging to the hoisting tackle is pulled by a machine worked with a crank.

13. Masonry is one of the primitive arts, and was carried to great perfection in ancient times. The pyramids of Egypt are supposed to have stood about three thousand years, and they will probably remain for centuries to come, monuments as well of the folly as of the power and industry of man. The temples and other magnificent structures of Greece and Rome, exhibit wonderful skill in masonry, and leave but little, if anything new, to be achieved in modern times.

THE BRICKMAKER.

1. Brick is a sort of artificial stone, made principally of argillaceous earths formed in moulds, dried in the sun, and burned with fire.

2. The earliest historical notice of bricks is found in the book of Genesis, where it is stated that the posterity of Noah undertook to build a city and a lofty tower of this material. Whether the bricks were really exposed to the action of fire, as the passage referred to seems to imply, or only dried in the sun, is an unsettled point. But Herodotus, who visited the spot many centuries afterwards, states that the bricks in the tower of Babylon were baked in furnaces.

3. It is evident, however, that the earliest bricks were commonly hardened in the sun; and, to give them the requisite degree of tenacity, chopped straw was mixed with the clay. The manufacture of such bricks was one of the tasks imposed upon the Israelites, during their servitude with the Egyptians.

4. The extreme dryness and heat of the climate in some of the eastern countries, rendered the application of fire dispensable; and there are structures of unburnt bricks still remaining, which were built two or three thousand years ago. Bricks both sun-dried and burned, were used by the Greeks and the Romans.

5. The walls of Babylon, some of the ancient structures of Egypt and Persia, the walls of Athens, the

rotunda of the Pantheon, the temple of Peace, and the Thermæ, or baths, at Rome, were all built of brick. The most common bricks among the Romans were seventeen inches long and eleven broad; a size, certainly, far preferable, as regards appearance, to those of modern manufacture.

6. In the United States, a great proportion of the edifices, particularly in the cities and towns, are constructed of bricks, which are usually manufactured in the vicinity of the place where they are to be used. The common clay, of which they are made, consists of a mixture of argillaceous earth and sand, with a little oxyde of iron, which causes them to turn red in burning. The material for bricks is dug up, and thrown into a large heap, late in the fall or in the winter, and exposed to the influence of the frost until spring.

7. The operation of making bricks is conducted very systematically; and, although every part of the work seems to be very simple, it requires considerable dexterity to perform it properly and to the best advantage. The workmen, in the yards about Philadelphia, are divided into *gangs* consisting of three men and a boy. The first is called the *temperer*, who tempers the material with water and mixes it with a spade; the second is called the *wheeler*, who conveys it on a barrow to a table, where it is formed in moulds by the *moulder*, whence it is carried to the *floor* by the boy, who is denominated the *off-bearer*.

8. The bricks are suffered to remain on the floor a day or two, or until they have become dry enough to be handled with safety. They are then removed and piled into a *hack*, under cover, in such a manner that the air may circulate freely between them. It is the business of the whole gang to remove the bricks from the floor, and also to place them in the kiln to be burned. In both cases, each one has his due proportion of labor to perform.

II.—K

9. The day's work of a gang, when the weather is favorable, is to make and pile in the hack a tale of bricks, which consists of 2332, or an even 2000. The former number is called a *long tale*, and the latter, a *short tale*. Considerable skill and much care are required in burning the bricks in a proper manner; too much fire would cause them to vitrify, and too little would leave them soft, and unfit for atmospheric exposure.

10. In many places, the clay is mixed or prepared for the moulder by driving round upon it a yoke of oxen, or by means of a simple machine, consisting of a beam, into which has been driven a great number of spokes. One end of this beam is confined in a central position, while the other is moved round in a sweep by animal power.

11. Machines have also been invented by the aid of which the clay may be both mixed and moulded; but these have been very little used. A machine, however, is often employed in pressing bricks which have been formed in the usual manner. The pressing is done after the bricks have become partially dry. Such bricks are employed in facing the walls of the better kinds of structures.

12. *Tiles.*—Tiles are plates used for covering roofs. They resemble bricks in their composition and mode of manufacture, and are shaped in such a manner that when placed upon a building, the edge of one tile receives that next to it, so that water cannot percolate between them. Tiles, both of burnt clay and marble, were used by the ancients; and the former continue to be employed in various parts of Europe. Flat tiles are used for floors in many countries, and especially in Italy.

THE BRICKLAYER.

1. The particular business of the bricklayer is to lay bricks in mortar or some other cement, so as to form one solid body; but he frequently constructs the foundations of buildings in rough stones, and, in some cities, he sets hewn stone in the superstructure. In the country, plastering is likewise connected with this business.

2. Bricklaying consists in placing one brick upon another in mortar, chiefly in the construction of walls, chimneys, and ovens. In connecting these materials, especially in walls, two methods are employed, one of which is called the *English bond*, and the other, the *Flemish bond*. In the former method, the bricks are most commonly of one quality, and are laid crosswise and lengthwise in alternate rows. The bricks which are laid across the wall are called *headers*, and those which are laid in the other direction are called *stretchers*. The brick-work of the Romans was of this kind, and so are the partition-walls of many modern brick edifices.

3. The bricks employed in the walls constructed according to the Flemish method, are of two, and frequently of three, qualities. Those placed in the front, or on the external surface, are manufactured with greater care, and, in some cases, are formed in a larger mould. A wall put up on this principle may be said to consist of two thin walls composed of stretchers, with occasional headers, to unite them together. The space between them, when the wall is thick, is filled in with the inferior bricks.

4. The inclosing walls of all brick edifices are erected on this plan, although they are thought to be more insecure than those constructed on the old English method. The reasons alleged for the preference, are its superior beauty, and a considerable sa-

ving in the most expensive kind of bricks. Greater security might be attained by the use of larger bricks, say sixteen inches in length, and wide and thick in proportion. Besides, an edifice constructed of well-made bricks of this size would be but little inferior in appearance to marble itself.

5. Most of the instruments used by the bricklayer are also employed by the stone-mason; and they have, therefore, been already mentioned. The particular method of laying bricks, in their various applications, can be learned by actual inspection in almost every village, city, or neighborhood, in our country, a more particular description of the bricklayer's operations is hence unnecessary.

6. Before closing this subject, however, it may be well to state that the chimney appears to be an invention comparatively modern, since the first certain notice we have of it is found in an inscription at Venice, in which it is stated that, in 1347, a great many chimneys were thrown down by an earthquake. It is conjectured that this valuable improvement originated in Italy, inasmuch as it was here that chimney-sweeping was first followed as a business.

7. Before the introduction of the chimney, it was customary to make the fire in a hole or pit in the centre or some other part of the floor, under an opening formed in the roof, which, in unfavorable weather, could be closed by a moveable covering. Among the Romans, the hearth or fire-place was located in the *atrium* or hall, and around it the *lares*, or household gods, were placed. To avoid being infested with smoke, they burned dry wood soaked in the lees of oil. In warming other apartments of the house, they used portable furnaces, in which were placed embers and burning coals.

8. It is said by Seneca, who flourished about the middle of the first century of the Christian era, that

in his time, a particular kind of pipes was invented, and affixed to the walls of buildings, through which heat from a subterranean furnace was made to circulate. By this means, the rooms were heated more equally. In the southern parts of Italy and Spain, there are still very few chimneys. The same may be said of many other countries, where the climate is pleasant or very warm.

9. Hollinshead, who wrote during the reign of Queen Elizabeth, thus describes the rudeness of the preceding generation in the arts of life: "There were very few chimneys even in capital towns: the fire was laid to the wall, and the smoke issued out at the roof, or door, or window. The houses were wattled, and plastered over with clay; and all the furniture and utensils were of wood. The people slept on straw pallets, with a log of wood for a pillow."

THE PLASTERER.

1. In modern practice, plastering occurs in many departments of architecture. It is more particularly applied to the ceilings and interior walls of buildings, and also in rough-casting on their exterior.

2. In plastering the interior parts of buildings, three coatings of mortar are commonly applied in succession. The mortar for the *first coat* is composed of about twelve parts of sand, six of lime, and three of hair, with a sufficient quantity of water to bring it to the proper consistence; that for the *second coat* contains a less proportion of lime and hair; and that for the *third coat* is composed exclusively of lime and water.

3. The mortar is applied directly to the solid wall, or to thin strips of wood called *laths*, which have been fastened with small nails to the joists, and other parts of the frame of the building. The tools with which the plasterer applies the mortar are *trowels* of different

sizes and shapes, and the *hawk*. The latter instrument is a board about a foot square, with a short handle projecting at right angles from the bottom.

4. In all well-finished rooms, cornices are run at the junction of the wall and ceiling. The materials of these cornices are lime, water, and plaster. The lime and water are first incorporated, and the plaster is added with an additional quantity of water, as it may be needed for immediate application. The composition is applied in a semifluid state, but the plaster causes it *to set*, or to become solid immediately. In the mean time, the workman applies to it, in a progressive manner, the edge of a solid piece of wood, in which an exact profile of the proposed cornice has been cut.

5. Ornaments of irregular shape are cast in moulds of wax or plaster of Paris, and these are formed on models of the proposed figures in clay. Such ornaments were formerly the productions of manual operations performed by ingenious men called *ornamental plasterers*. The casts are all made of the purest plaster; and, after having been polished, they are fastened to the proper place with the same substance saturated with water.

6. The branch of this business called *rough-casting*, consists in applying mortar to the exterior walls of houses. The mode in which the work is performed varies but little from that adopted in plastering the walls of apartments. It, however, requires only two coats of the cement; and, when these have been applied, the surface is marked off in imitation of masonry. It is likewise sometimes colored, that it may resemble marble or some other stone.

7. The cement is commonly made of *sharp sand* and lime; but sometimes a kind of argillaceous stone, calcined in kilns and afterwards reduced to powder by mechanical means, makes a part of the composition.

The qualities of this material were first discovered by a Mr. Parker, who obtained letters patent for this application of it, in England, in 1796; hence it has been called *Parker's cement.*

THE SLATER.

1. SLATE stone is valuable for the property of splitting in one direction, so as to afford fragments of a sufficient size and thinness to answer several purposes, but especially for covering houses and for writing slates. The best slates are those which are even and compact, and which absorb the least water.

2. The slates used in the United States, are obtained either from our own quarries, of which there are several, or from those of Wales, in the county of Caernarvonshire. The stone is quarried in masses, which are afterwards split into pieces of suitable thinness. These are trimmed to an oblong figure by means of a knife and a steel edge, which act upon the slate much in the manner of a large pair of shears.

3. As it is impossible to dress all the slates to the same size without much waste of material, those engaged in their manufacture have introduced several sizes, the smallest of which are made of the fragments of the larger kinds. These are designated by names known to the trade, and to those practically conversant with the art of building.

4. The slates, when brought to market, especially those from Wales, require additional dressing to fit them for use. The manner of applying them to roofs differs but little from that employed in putting on shingles, as they are lapped over each other in the same way, and confined to their place by means of nails of a similar kind. The nails, however, have a broader head, and are somewhat larger, varying in size to suit the dimensions of the slate. The holes in the slate for the nails are made with a steel point at-

tached to the slater's hammer, or to his knife, technically called a *saix*.

5. Slates are preferable to shingles on account of their durability, and, in a majority of situations, for their fire-proof quality. They, however, are objectionable on account of their weight and expensiveness, and are therefore beginning to be superseded in this country by sheets of zinc, and by those of iron coated with tin. Copper and lead are also used for roofs, but the metals just mentioned are beginning to exclude them altogether.

6. A serious objection to metal roofs has been their liability to crack, caused by the contraction and expansion of the material, in consequence of variations in the temperature of the weather; but a particular method of putting the sheets together has been lately devised, which appears to obviate the difficulty. Tiles are not used in this country, although in Europe they are very common.

THE PAINTER, AND THE GLAZIER.

THE HOUSE AND SIGN PAINTER.

1. THE painting which is the subject of this article relates to forming letters and sometimes ornamental and significant figures on signs, as well as to the application of paints to houses and other structures, for the purpose of improving their appearance, and of preserving them from the influence of the atmosphere and other destructive agents.

2. The substances capable of being employed by the house and sign painter, comprise a great variety of articles, derived from the mineral, vegetable, and animal kingdoms; but he ordinarily confines his selection to but few, among which are white lead, litharge, Spanish brown, yellow ochre, chrome yellow,

red ochre, terra di sienna, lampblack, verdigris, linseed-oil, spirits of turpentine, and gold-leaf.

3. White lead and litharge are manufactured in great quantities at chemical works, sometimes established for the express purpose of making these and some other preparations of lead. The substances of which we are now speaking, are produced in the following manner: the lead, in form of a continued sheet, about three feet long, six inches wide, and one line in thickness, is wound spirally up in such a manner, that the coils may stand about half an inch apart.

4. The metal in this form is placed vertically in earthen vessels, at the bottom of which is some strong vinegar. These vessels, being placed in sand, horse manure, or tan, are exposed to a gentle heat, which causes the gradual evaporation of the vinegar. The vapor thus produced, assisted by the oxygen which is present, converts the exposed surface into a carbonate of lead, the substance known as white lead, or ceruse.

5. The corrosion of one of these sheets occupies from three to six weeks, during which time it is repeatedly uncoiled and scraped. Litharge, or flake white, is nothing more than the densest and thickest scales produced in the manner just described. It can be obtained in a pure state from the dealers in paints, whereas the white lead of commerce is most commonly adulterated with chalk.

6. Spanish brown, yellow ochre, and terra di sienna, are earths impregnated with iron in different degrees of oxydation. Red ochre is yellow ochre burned. Chrome yellow is extensively manufactured in Baltimore, from the chromate of iron, found near that city. In chemical phraseology, the manufactured article is the chromate of lead, since the chromate is

separated from the iron by the aid of a solution of the nitrate or acetate of lead.

7. Linseed-oil is obtained from flax-seed by pressure. It is afterwards filtered, and then suffered to remain at rest, to precipitate and clarify. This oil improves in quality by keeping, as it becomes, in a few years, as transparent as water. In this state, it is employed in the finest painting.

8. Before the oil is used, it is commonly boiled with a small quantity of litharge and red lead, to cause it to dry rapidly, after the paint has been applied. During the boiling, the scum is removed as fast as it rises, and this is mixed with inferior paints of a dark color. Linseed-oil, thus prepared, is vended by dealers in paints, under the name of boiled oil.

9. Spirits of turpentine is produced by distilling with water the resinous juice or sap of several species of the pine. The residuum, after distillation, is the turpentine of commerce. Spirits of turpentine is mixed with paints, to cause them to dry with rapidity. Like oil, it improves with age, and it is sold in the same manner by the common wine measure.

10. White lead, and several other principal paints, are purchased in their crude condition, and reduced to a state of minute division in paint-mills. They are afterwards mixed with boiled oil, and put up in kegs of different sizes for sale. Many articles, however, are pulverized, and sold in a dry state. The preparation of paints is commonly a distinct business, and very few painters seem to be acquainted with the mode in which it is performed.

11. In mixing colors for house and sign painting, white lead forms the basis of all the ingredients. This the color preparer, or the painter himself, modifies and changes by the addition of coloring materials, until it is tinged with the proposed hue. The pigments derived from vegetable bodies, produce, when

first applied to surfaces, a brilliant effect; but they cannot long resist the combined influence of air and light, while the mineral colors, in the same exposure, remain unchanged.

12. Painters, in the execution of their work, commonly lay on three coats of paint. In communicating a white, the two first coats are composed of white lead and oil; and in the last, spirits of turpentine is substituted for the oil, for the inside work. For the outside of buildings, especially in warm and dry climates, this liquid is inapplicable, since it causes the paint to crack and flake off. It is, however, frequently used, when the painter is compelled to do his work at too low a rate, or when he is regardless of his reputation.

13. For other colors, the composition for the different coats is the same, except for the two last, in which other coloring substances are added to the materials just mentioned, to give the proposed hue. The tools for painting houses are few in number, and consist chiefly of brushes of different sizes, made of hog's bristles.

14. *Graining* is understood, among painters, to be the imitation of the different species of scarce woods used for the best articles of furniture. But the manner in which this kind of work is executed can be hardly gathered from a concise description, although it may be easily learned from a practical exhibition of the process by a painter.

15. *Ornamental painting* embraces the execution of friezes and other decorative parts of architecture on walls and ceilings. The ornaments are drawn in outline with a black-lead pencil, and then painted and shaded, to give the proper effect. Some embellishments of this kind are executed in gold-leaf, in the same manner with gold letters on signs. This kind of work is called *gilding in oil*.

16. Painting in oil, as applied to the execution of designs, seems to have been invented, or at least to have been brought into notice, in the early part of the fifteenth century, by John Van Eyck, of Flanders. Before this time, house-painting, so far as the exterior was concerned, could have been but little, if at all, practised.

17. One profitable branch of common painting is that of painting and lettering signs. In performing this kind of work, the sign is first covered with two or three uniform coats of paint. The letters are next slightly sketched with chalk or a lead-pencil, and then formed in colors with a camels'-hair brush. When the letters are to be gilt, the process, so far, is precisely the same. The leaf is laid upon the letters, while the paint is in a tenacious state, and is suffered to remain untouched, until the oil has become dry, after which the superfluous gold is removed. The whole is then covered with an oil varnish, which, in plain lettering, completes the operation.

THE GLAZIER.

1. GLAZING, as practised in this country, consists chiefly in setting panes of glass in window-sashes. In the performance of this operation, the glazier first fits the panes to the sash by cutting away, if necessary, a part of the latter with a chisel; he then fastens the glass slightly with little pieces of tin, which have been cut to a triangular shape; and, lastly, he applies *putty* at their junction with the sash, and by this means confines them firmly and permanently to their place. The putty is made of linseed-oil and whiting. The latter of these materials is chalk cleared of its grosser impurities, and ground in a color-mill.

2. Plain glazing is so simple, that no person need serve an apprenticeship to learn it; and there are but few who confine their attention to this business exclu-

sively. It is commonly connected with some other of greater difficulty, such as that of the carpenter and joiner, or house and sign painter, but with the latter more frequently than any other.

3. When the glass, as received from the manufacturer, may not be of the size and shape required for a proposed application, the panes are cut by means of a diamond fixed in lead, and secured by a ferrule of brass, which is fastened to a small cylindrical handle of hard wood. This instrument is used, in conjunction with a straight edge, like a pencil in ruling lines on paper for writing. The glass is afterwards broken in the direction of the fracture, by a slight pressure downwards.

4. Although glass windows seem to us to be indispensable to comfort, yet glass had been manufactured many centuries in considerable perfection, before it was applied to this purpose. The houses in oriental countries had commonly no windows in front, and those on the other sides were provided with curtains, or with a moveable trellis-work in summer, and in winter with oiled paper.

5. In Rome and other cities of the empire, thin leaves of a certain kind of stone called *lapis specularis* were used. Windows of this material, however, were employed only in the principal apartments of great houses, in gardens, sedans, and the like. Paper made of the Egyptian papyrus, linen cloth, thin plates of marble, agate, and horn, seem likewise to have been used.

6. The first certain information we have of the employment of glass panes in windows, is found in the writings of Gregory of Tours, who flourished in the last quarter of the sixth century. This prelate states that the churches were furnished with windows of colored glass, in the fourth century after Christ. The oldest glass windows now in existence were of

the twelfth century, and are in the Church of St. Denis, the most ancient edifice of this description in France.

7. Æneas Sylvius accounted it one of the most striking instances of splendor which he met with in Vienna, in 1458, that most of the houses had glass windows. In France, all the churches had these conveniences in the sixteenth century, although there were but few in private dwellings. Talc, isinglass, plates of white horn, oiled paper, and thinly shaved leather, were used instead of glass. A similar state of things prevailed in England.

8. The glass used for the windows of churches and other public buildings, after the fourth century, was very commonly intrinsically colored or superficially painted. Painting on glass had its origin in the third century, and at first it consisted in the mere arrangement of small pieces of glass of different colors in some sort of symmetry, and constituted a kind of mosaic-work.

9. Afterwards, when more regular designs came to be attempted, such as the human figure, the whole address of the artist went no farther than drawing the outlines of the objects in black on glass resembling in color the subjects to be represented. The art, in this state of advancement, was spread over a great part of Europe.

10. About the beginning of the fifteenth century, a method of fixing metallic colors in glass by means of heat was discovered, and from this the art derived great advantages. It flourished most during the fifteenth and sixteenth centuries; but it declined in the following age, and in the eighteenth century it was very little practised in any country. It has, however, been partially revived, of late, in Germany. A very good specimen of this kind of painting, as well as of colored glass, may be seen in St. John's Church, in Philadelphia.

THE TURNER.

1. Turning is a very useful art, by which a great variety of articles are almost exclusively manufactured. Besides this, it constitutes a considerable part of the operations of several trades and occupations, such as the chairmaker, machinist, cabinet-maker, brass-founder, &c., since every substance of a solid nature can be submitted to the process.

2. Turning is performed in a *lathe*, an apparatus constructed in various ways, according to the particular purposes to which it is to be applied, although, in all cases, the general principle of its operation is the same. The kind represented in the above picture, is used for plain or circular turning in wood. On examination, it will be perceived, that two wheels of different sizes make essential parts of it. On the extend-

ed axle of the smaller one, is fastened the piece to be turned; and immediately in front of this is the *rest*, on which the cutting instrument is supported during the performance of the operation.

3. When the material to be turned is wood, it is commonly cut to the proper length with a saw, and brought to a form approaching to the cylindrical by means of an axe or drawing-knife. It is next fastened in the lathe. This is done by different means, varying according to the particular form of the thing to be turned. In plain circular turning, as applied to bed-posts, legs of tables, and rounds for chairs, the piece is supported at each end. That at the left hand is driven upon a piece of steel, which has been screwed upon the extended axle of the small wheel; and the other end is fixed upon a steel point, placed in an upright moveable piece called a *puppet-head*.

4. In case the wood is to be turned on the inside, as in making a bowl, cup, or mortar, the piece is supported altogether at one end, by means of a hollow cylinder of wood, brass, or iron, called a *chuck*, which receives it on one side, and on the other is screwed upon the end of the axle. The axle is sometimes called the *mandril*, and any extension of it, by means of a piece added to it for a centre, on which anything may be turned which will admit of a hole through it, is denominated an *arbor*.

5. The tools used in turning wood and ivory, are *gouges* and *chisels* of different sizes and shapes. In using these, they are placed upon the *rest*, and brought in contact with the revolving material of the proposed figure. The gouge is employed in cutting away the rough exterior, and the chisel, in producing a still further reduction, and a greater smoothness of surface.

6. In working in very hard wood and in ivory, the *grooving tool*, a sharp pointed instrument somewhat similar to the graver, is used in the first part of the

operation; and by this the grain of the substance is cut into contiguous grooves, and prepared for an easy reduction by the chisel. The instruments for turning metals are numerous, but they differ in some respects from those for cutting wood.

7. In almost every kind of turning, a tool called the *calipers* is necessary for measuring the diameters of the work. In its form, it bears some resemblance to the compasses or dividers. One or both of the legs, however, are curved; and one kind of this instrument has four legs, two curved, or two straight, at each end, with a pivot in the centre, on which it is opened and shut. The former of these is employed in measuring the dimensions of outside work, and the latter, for that on the inside. This kind is called the *in-and-out* calipers; and it is especially useful in turning a cylinder, or pin, which shall exactly fit an internal cylinder already made, and *vice versâ*.

8. There is but little difference in the management of turning different substances. The principal thing to be attended to is to adapt the velocity of the motion to the nature of the material; thus wood will work best with the greatest velocity that can be given to it. Brass should have a motion about half as quick as wood, and iron and steel still less; for, in operating on metallic substances, the tool is liable to become hot, and lose its temper; besides which, a certain time is requisite for the act of cutting to take place.

9. When compared with many other mechanical operations, the art of turning may be considered as perfect in its accuracy and expedition. The lathe is, therefore, resorted to for the performance of every work of which it is capable; nor is its use confined to the production of forms perfectly cylindrical, for it can be easily made to produce figures of irregular shape, such as lasts, gunstocks, &c.

10. The lathe was well known to the Greeks and

Romans, as well as to many other nations of antiquity. Diodorus Siculus, who wrote in the time of Julius Cæsar and Augustus, says that it was invented by one Talus, a nephe of Dædalus. Pliny ascribes it to Theodore, of Samos, and mentions one Thericles, who had rendered himself very famous by his dexterity in managing the lathe. The Greek and Latin authors frequently mention this instrument; and, among the ancients, it was customary to express the accuracy and nicety of a thing by saying, it was formed in a lathe.

THE CABINET-MAKER, AND THE UPHOLSTERER.

THE CABINET-MAKER.

1. It is the business of the cabinet-maker to manufacture particular kinds of household furniture, such as tables, stands, bureaus, sideboards, desks, bookcases, sofas, bedsteads, &c., as well as a certain description of chairs made of mahogany and maple. Many of the operations of this business are similar to those of the carpenter and joiner, although they require to be conducted with greater nicety and exactness.

2. The qualifications of a finished cabinet-maker are numerous and of difficult acquisition; so that they are seldom concentrated in any single individual. He requires not only a correct taste, but also a knowledge of drawing, architecture, and mechanics, besides the abilities of a good practical workman.

3. A knowledge of drawing is especially useful in designing new articles of furniture, or in improving the form of those which have been already introduced. It also enables the artist to determine with accuracy what would be the general effect of furniture, were different pieces of it placed in any proposed apartment; and, combined with architectural knowledge, it enables him to adapt the style of his wares to that of the building for which they may be designed.

4. In general, the principles of this business are fixed, so far as relates to the mode of operating in the execution of the work; yet continual changes are made in the form and construction of its various articles, so as to keep pace with the advancement of correct taste, or with the caprices of fashion. In fact, the shapes of furniture are almost as changeable as those of female dress; and this causes many expensive pieces to fall into disuse, while others are introduced, which, for a time, are considered indispensable to comfort, and which in turn enjoy but a temporary favor.

5. The cabinet-maker uses various kinds of wood in the manufacture of his wares; but those which are most frequently employed in the United States are pine, maple, poplar, cherry, black walnut, white oak, beach, mahogany, and rose, all of which are abundant in this country, except the last two. Mahogany is brought in great quantities from the West Indies and South America; rose-wood is obtained chiefly from the West Indies and Brazil, although it was first introduced into notice from the island of Cyprus.

6. The applicability of mahogany to the manufacture of cabinet-ware, was accidentally discovered in London, about the year 1724. A physician, named Gibbons, received a present of some of the planks from his brother, a sea-captain, who had brought them from the West Indies, chiefly as ballast. The

doctor was, at that time, erecting a house, and, supposing them to be adapted to the purposes of building, gave them to his workmen, who, on trial, rejected them as being too hard to be wrought with their tools.

7. A cabinet-maker was next employed to make a candle-box of some of it, and he also complained of the hardness of the timber; but, when the box was finished, it outshone in beauty all the doctor's other furniture. He then required a bureau to be made of the same kind of material; and this, having been finished, became the subject of exhibition to his friends, as a piece of remarkable beauty. The wood was immediately taken into general favor, and it soon became an article of merchandise of considerable importance.

8. In giving the reader a view of the operative part of this business, we have selected the bureau as affording the best means of illustration. The material which composes the frame and drawers of this piece of furniture, is commonly some kind of soft wood, such as pine or poplar; and this is faced with thin layers of mahogany in those parts which are to be exposed to view.

9. The materials for the frame and drawers are first marked out, and the several pieces reduced to the form and dimensions required, with planes and other instruments. Thin pieces of mahogany are firmly fixed to the surfaces which require them. This part of the work is called *veneering*. The workman prepares the surface of the soft wood for the *veneer*, by cutting it into small contiguous grooves by means of a small plane, the cutting edge of which is full of little notches and teeth.

10. Melted glue having been spread upon both surfaces with a brush, the parts are placed in contact, and firmly pressed together by means of *hand-screws*. Before the screws are applied. the surface of the ve-

neer is covered with a piece of heated board, termed, in this application, a *caul*. One piece of this kind commonly serves a veneer on each side of it at the same time.

11. The mahogany thus attached to the softer wood, is afterwards wrought with the *toothed-plane*, and others of the common kind. It is then scraped with a flat piece of steel, having edges which act upon the surface in the same manner as pieces of broken panes of glass. The polishing is finished, so far as it is carried at this stage of the process, by the use of sand-paper.

12. The several pieces which compose the frame of the bureau are put together with the joint called *mortice* and *tenon;* and those which form the four sides of the drawers, with that called *dove-tail*. The bottom is united to the sides on the right and left, and sometimes in front, by the *groove-and-tongue*, and its rear edge is fastened with a few nails. The *bearers* of the drawers are fastened on by means of nails.

13. The joints are made to fit not only by the accuracy of the work, but by the application of glue previous to the union of the parts; this is especially the case with the mortice and tenon. The back of the bureau is composed of some cheap wood, such as pine or poplar; but the panel at each end is most commonly plain mahogany through its entire thickness.

14. The parts which are to be exposed to view are next to be varnished and polished. The material for the former purpose is called *copal varnish,* because one of the principal ingredients in it is a kind of gum called copal, which is obtained from various parts of South America. This kind of varnish is made by melting the gum with an equal quantity of linseed-oil and spirits of turpentine or alcohol.

15. To give the work a complete finish, four coats

of varnish are successively applied; in addition to these, a particular kind of treatment is used after laying on and drying each coat. After the application of the first coat, the surface is rubbed with a piece of wood of convenient form; after the second, with sandpaper and pulverized pumice-stone; after the third, with pumice-stone again; and after the fourth, with very finely powdered pumice-stone and rotten-stone. A little linseed-oil is next applied, and the whole process is finished by rubbing the surface with the hand charged with flour.

16. Some parts of several pieces of furniture are turned in the lathe; and, in large cities, this part of the work is performed by professed turners. The veneering of certain kinds of work of a cylindrical form is, also, in some cases, a distinct business; but, in places distant from large cities, the whole work is commonly performed by the cabinet-maker himself.

17. Mahogany is brought to market in logs hewn to a square form; and persons who deal in it, commonly purchase it in large quantities, and cause it to be sawn into pieces of suitable dimensions for sale. Formerly, and in some cases at present, slabs were sawn into thin pieces for veneering by hand; but, within a few years, a more expeditious method, by the circular saw, has been adopted. In performing the operation by this means, the slab is placed upon its edge, and shoved along against the teeth of the rapidly-revolving saw. It is kept in the proper position by holding the right side of it firmly against an upright plank, called the *rest*.

18. Mahogany is either *plain, mottled,* or *crotched;* nevertheless, the different kinds expressed by these terms are met with in the same tree. The variegated kinds are found at or near the joining of the limbs to the trunk; and these are used almost exclusively for veneering. The plain sort is employed for more

common purposes, and in those parts of furniture required to be less splendid in appearance. It may be well to remark, also, that plain mahogany is often veneered, as well as the softer woods. Black walnut, white oak, rose, and several other woods, are likewise used for veneering, although not so much as mahogany. Our native woods will be hereafter more used in this way, since mahogany is becoming scarce.

19. In Europe, particularly in England, the business of the cabinet-maker is commonly united with that of the upholsterer; and this is sometimes the case in the United States. All, however, who make sofas and chairs, intrude enough upon the latter business to cover and stuff them; or they employ a journeyman upholsterer to perform this part of the work.

THE UPHOLSTERER.

1. THE upholsterer makes beds, sacking-bottoms, mattresses, cushions, curtains for windows and beds, and cuts out, sews together, and fastens down, carpets. One branch of his business, also, consists in covering or lining and stuffing sofas, and particular kinds of chairs, the frames of which are made by cabinet-makers and fancy chair-makers.

2. Beds are stuffed with the feathers of geese and ducks. The sack which contains them, when in use, is called a *tick*, and the striped stuff of which it is composed, is called *ticking*. The feathers used by the upholsterer, are purchased from the feather-merchants, who in turn procure them from country merchants and pedlers. The dealer in feathers also employs travelling agents to collect them in different parts of the country.

3. Beds and pillows are also made of down obtained from the nests of the eider-duck, which is found in the northern parts of Europe and America, above lat.

itude 45°. Eider-down is worth about two dollars per pound, and five or six times that quantity is sufficient for a bed of common size.

4. Mattresses are made of curled hair, moss, shavings of ratan, flock, straw, corn-husks, and cat-tail flag. The hair most employed for this purpose grows upon the tails of cattle, and upon the manes and tails of horses. It is purchased, in its natural state, from tanners, by persons who make it a business to prepare it for use. The last process of the preparation consists in twisting it into a kind of rope. These ropes are picked to pieces by the upholsterer, and the hair, in its curled and elastic state, is applied to stuffing mattresses, cushions, chairs, and sofas.

5. Moss is obtained from the Southern states of our Union, where it is found in great abundance, and of a good quality. Flock is made by reducing to a degree of fineness, by machinery, coarse tags of wool, pieces of woollen cloth, old stockings, and other woollen offals of little or no value in any other application Of all the materials for stuffing upholstery, hair is much the best, and, although it costs more in its original purchase, it is much cheaper in the end.

6. In making and putting up window and bed curtains, considerable taste is required to insure success. A knowledge of drawing is particularly useful here, in improving the taste, as well as in exhibiting to customers the prevailing fashions, or any changes which may be proposed. The trimmings consist chiefly of tassels, fringes, and gilded or brass fixtures.

7. We have not space for a particular description of the manner in which any of the operations of the upholsterer are performed; nor is this necessary, since the work itself, in almost every specimen of it, affords obvious indications of the manner of its execution. We will merely remark, that a great proportion of it is performed by females.

8. In the first ages of the world, it was the universal practice to sleep upon the skins of beasts, and this is still the custom among the savage nations of the present day. The Greeks and the Romans, in the early part of their history, slept in this manner, and so did the common people of some parts of Germany, even until modern times.

9. The first advancement from the use of skins was the substitution of rushes, heath, or straw, which was primarily strewed loosely on the ground or floor, and finally confined with ticking; and these and similar materials are still used by the poor in various parts of the world. So late as the close of the thirteenth century, the royal family of England slept on beds made of straw.

10. During the civilized periods of antiquity, the wealthy commonly filled their beds with feathers. After the Romans had become luxurious, they used several kinds of beds, among which were the *lectus cubicularis*, or chamber bed, whereon they slept; the *lectus discubitorius*, or table bed, whereon they ate; and the *lectus lucubratorius*, on which they studied.

11. The Romans adopted the Eastern fashion of reclining at their meals, at the close of the second Punic war, about 200 years before Christ, when Scipio Africanus brought some little beds from Carthage, which were thence called *Punicani*. These beds were low, made of wood, covered with leather, and stuffed with hay or straw. Before this time, they sat down to eat on plain wooden benches, in imitation of the heroes of Homer, or after the manner of the Cretans and Lacedæmonians.

12. From the greatest simplicity, the Romans at length carried their supping beds to the most surprising magnificence. The bedsteads were sometimes made of gold or silver, and very commonly of wood, adorned with plates of these metals or with tortoise

shell. On the couch was laid a mattress or quilt, stuffed with feathers or wool.

13. Three persons commonly occupied one couch. They lay with the upper part of the body reclined on the left arm, the head a little raised, the back supported by cushions, and the limbs stretched out at full length or a little bent. The feet of the first were placed behind the back of the second, and his feet behind the back of the third. Reclining at meals was customary in Asia, in the time of our Savior, as is clearly shown in John, xiii., 23 and 25, and this rendered it convenient for Mary to anoint the feet of Jesus, while at the table.

14. The Romans, during the republic, made their tables of a square form, and on three sides of it was placed a couch; but, under the emperors, a long couch of a semicircular form having been introduced, the table was made of a similar shape to conform to it. In either case, one side was left empty, to admit of the approach of the servants.

15. We have no certain evidence that carpets were known in the civilized periods of antiquity. They appear to have originated in Persia, at a time comparatively modern, and to have spread in a gradual manner towards the West. They were unknown in England in the reign of Elizabeth; for it was then the fashion to strew the floor with hay and rushes. Even the presence-chamber of this princess was covered in this manner. The manufacture of carpets was not commenced in England, until the year 1750. They are now extensively manufactured in the United States

THE CHAIR-MAKER.

1. THE chair was invented at so early a period, that its origin cannot now be ascertained. It was used by all the civilized nations of antiquity; and some of their patterns for this species of furniture have been revived, with some modifications, in modern times; for example, a stool for sitting at the piano, now called the X, is the lower part of a chair used in the Roman empire near two thousand years ago. The seat and back were stuffed with some soft elastic substance.

2. The seats used by the barbarous conquerors of the Roman empire, hardly deserve the name of chairs, as they commonly consisted of little or nothing more than a stool with three or four legs. Even the great Alfred, who swayed the sceptre of England in the

latter part of the ninth century, possessed nothing approaching nearer to a chair than a three-legged stool made of oak timber. This species of seat was at length improved into a chair by the addition of another leg and a back.

3. The next step in the art of chair-making was to cover the seats with cloth, and to stuff them with some kind of wadding. The material of which the frames were made was oak; and for a long period, they were exceedingly heavy and inconvenient. The armed-chair is said to have been contrived by an alderman of Cripplegate. Such chairs, however, were in use among the ancient Greeks and Romans.

4. Our old-fashioned chair, with four upright posts, several horizontal rounds and slats, together with wooden splints or flags for the bottom, is comparatively modern, although it is impossible to state the period of its introduction. Very few of any other kind were used in the United States, until near the beginning of the present century.

5. The Windsor chair seems to have been first used for a rural seat in the grounds about Windsor castle, England; whence its name. It was originally constructed of round wood, with the bark on; but the chair-makers soon began to make them of turned wood, for the common purposes of house-keeping. We cannot learn that any were made in this country before the close of the revolution, in 1783.

6. A great proportion of the chair-maker's stuff is brought to the proper form by means of the lathe; and this machine is used for this purpose in every practicable case; but this part of the work is not performed in the cities, since it is found to be less expensive and more convenient, to purchase the timber turned in the country. Slats for the back, bent to the proper shape, are also obtained from the same source.

7. The Windsor chair is varied in its construction

and finish, in some particulars; but, in all cases, it has a seat made of thick plank of cypress, bass, or some other soft wood. The slats, when employed, are also made of the same wood, or of soft maple. The parts which are turned, are commonly of the wood last mentioned.

8. In constructing chairs from these materials, the workman undertakes several at a time, say from one to two or three dozens. We may suppose, as is frequently the case, that he first cuts up a quantity of planks to the proper size for the seats, and reduces them to the proposed form and smoothness by means of the drawing-knife, adze, spoke-shaves, and sand-paper. He next cuts the various pieces which are to compose the frame, to the proper length, turns the ends of those which need it, to make the joint, and bores the requisite holes with a *bit*. In putting the parts together, the joints are made to fit very closely, and their union is rendered permanent by means of glue.

9. The chairs are next covered with three coats of paint, and with two coats of copal or some other kind of varnish; and this, for plain work, completes the whole process of the manufacture. But, when they are to be ornamented, gold or copper leaf or bronze is put on before the application of the last coat of varnish. The bronze used by painters, is finely pulverized copper, tin, or zinc.

10. The *ornamenter* uses paper patterns, which he applies to the surface to be ornamented, to guide him in the execution of his work. The powder is laid on with a camel's-hair brush, or with a piece of raw cotton. Light and shade are produced by a proper distribution of the powder, or by paint of a dark colour. The bronze is made to adhere by means of *size*, which has been previously laid on.

11. Several other kinds of chairs are, also, made

by the common chair-maker; and the frames, or some parts of them, are sawn out of planks with a narrow-bladed saw, which can be easily guided upon the line of any pattern. The principal parts of the frame are commonly put together with the mortice and tenon; and the bottoms are composed of cane, flags, or a peculiar kind of rush. The cane is likewise used in the backs of chairs, especially in those having rockers.

12. The manufacture of mahogany chairs with stuffed seats, sometimes constitutes a distinct branch of business; at other times, it is connected with that of making sofas; and again, with cabinet-making in general. It is generally supposed, that rockers were first applied to chairs in this country, but at what time or by whom, it cannot be determined

THE CARVER, AND THE GILDER.

THE CARVER.

1. CARVING, in its widest sense, is the art of forming figures in various hard substances by means of some cutting instruments, such as a chisel or graver; but, in the restricted sense in which the term is generally applied, it has reference to the production of figures in wood.

2. Carving in wood, in all countries where it has been practised, has ever preceded sculpture, or carving in stone. It is, therefore, an art of the highest antiquity; and, although the same with sculpture in some of its applications, yet it differs from it somewhat in the mode of execution, according with the nature of the material.

3. The art of carving is very extensive in its appli

cation, being used in the decorative parts of architecture, both civil and naval, and likewise in ornamenting cabinet-ware, as well as in forming patterns for casting in metals, particularly in iron and brass. The Gothic style of architecture is peculiarly rich in carved work; and the productions of some ages are more so than those of others.

4. The style of Louis the Fourteenth, of France, so called because practised in his reign, was more overloaded with ornament than any other. A lighter and more beautiful style succeeded, which is still employed for some purposes; but generally the chaste and simple line of Grecian ornament now prevails.

5. In executing any proposed work, a drawing is first made on paper, commonly with a lead-pencil. The part of the paper not embraced in the outline is then cut away, and the remaining portion is laid upon the surface of the wood. The outlines are next drawn on the wood, by moving the pencil around those on the paper. The design having been thus transferred, the superfluous portions of the wood are cut away with carving tools, of which there is a considerable variety of both size and form. The tools are driven with a mallet or with the palm of the hand, but in most cases with the latter.

6. A capacity for designing, and a knowledge of drawing and modelling, are particularly necessary to make a finished carver. Without these qualifications, at least in some degree, one may be a mechanic, but not an artist. The subject most difficult of execution, is the human figure, and in producing it with accuracy, the same qualifications in the artist are required, and the same general process is pursued, as in producing it in marble.

THE GILDER.

1. CARVING and gilding are, in most cases, ostensibly united as one business, although in fact they are branches of manufacture totally distinct. The gilder, therefore, who writes over his door, "Carver and Gilder," seldom has any practical knowledge of carving. For every thing in this line of work, he is dependent on the carver, who commonly pursues his business in a private way.

2. The operation of gilding, as performed by those whose business is now under consideration, is executed chiefly on wood. It is employed most frequently for picture and looking-glass frames, and for upholstery fixtures. It is a mechanical process, and consists in applying gold-leaf to surfaces, in such a manner as to adhere with tenacity.

3. Before the application of the metal, a tedious process must be performed, by way of preparation. The surface to be gilded is successively covered with from five to seven coats of glutinous size, made by boiling scraps of parchment in water, with the addition of a little whiting. The average thickness of the coat thus produced, is about one-sixteenth of an inch.

4. The surface is next rubbed with freestone and pumice stone, of a shape corresponding with the pattern of the frame, while a small quantity of water is occasionally applied, to increase their effects. After this, the sizing is rendered still smoother, by friction with sand-paper. This surface is then covered with three coats of *burnished gold size*, which is composed of English pipe clay, venison suet, and French bole, or red chalk, mixed in a suitable quantity of weak parchment size. The preparation is completed by rubbing the surface with worn sand-paper, by washing it in water with a sponge, and by rubbing it with a piece of cloth.

5. The leaf is laid on with a broad, but thin brush, called a *tip*. Before the gold is applied, however, the surface is well wet with alcohol and water. When dry, the parts designed to be bright, are burnished with a polished agate or flint. In the best kind of work, a second coat of the leaf is required. In gilding irregular surfaces, such as the ornaments at the corners of frames, a size made of linseed-oil, white lead, yellow ochre, and japan, is laid on a few hours before the application of the leaf. This is called *gilding in oil.*

6. The ornaments on the frames are cast in moulds, and are made of a composition of glue, whiting, rosin, turpentine, and Burgundy pitch. The moulds are taken from patterns, originally executed by the carver.

THE COOPER.

1. THE cooper manufactures casks, tubs, pails, and various other articles for domestic use, as well as vessels for containing all kinds of liquids and merchandise of a dry nature. He also applies hoops to boxes which are to be transported, with their valuable contents, to a distance from the cities.

2. The productions of this art being of prime necessity, the trade must have been exercised at a very early period. Roman writers on rural economy speak of the existence of its productions more than two thousand years ago; nevertheless they are still unknown in some countries, and there the inhabitants keep or carry liquids in skins daubed over with pitch.

3. Bottles of this kind were used, more or less, in all parts of the Roman empire, in the days of our Savior; and to such he alluded, when speaking of put-

ting new wine into old bottles. Earthen vessels of various dimensions, were also in extensive use at the same time. The custom of keeping wine in such vessels, is still common in the southern parts of Europe. Pliny accords to the Piedmontese the merit of introducing casks. In his time, they were daubed with pitch.

4. Cedar and oak are the woods chiefly employed as materials in this business; and the persons who carry it on, as well as journeymen, confine their attention to the production of wares from one or the other of these woods; hence the division of the workmen into *cedar coopers* and *oak coopers*.

5. It is not always the case, however, that every cooper executes all kinds of work belonging to either one of these divisions of the trade; but this is not because there is any peculiar difficulty attending any part of the business, but because some particular kind of coopering is required in preference to others; for example, in some places, flour barrels are the casks most needed; in others, those for sugar, tobacco, pearlash, or some kind of spirits.

6. In illustrating the general operations of this business, we will describe the process of making a tub. The timber is first cut to the proper length with the kind of saw used in the cities for cutting fire-wood. It is next split into pieces with a *frow*, the curvature of which corresponds, at least with some degree of exactness, to that of the proposed vessel. The several pieces are then shaved on the edges with a straight *drawing-knife*, on the inside with one of a concave form, and on the outside with one of corresponding convexity.

7. After this, they are jointed on a long plane, which is placed with its face upwards, in an inclined position. The workman is guided in giving the proper angle to the surface cut with the plane, by a wood-

an gauge of peculiar form. The staves, having been thus prepared, are set up in a *truss-hoop;* and after this has been driven down, one or two others which are to remain are put on. The outside is then made smooth with a onvex drawing-knife, and the inside with a smoothing-plane, the edge of which is circular, to correspond with the form of the surface. The inside of small wooden vessels is generally made smooth with a crooked drawing-knife.

8. The staves are now sawn off to a uniform length at the bottom, and a groove is cut for the insertion of the bottom. The latter operation is performed by means of a cutting instrument fixed in a kind of gauge. The several pieces to compose the bottom are brought to the proper form and smoothness with a straight drawing-knife; and, having been slightly fastened together by wooden pins, the whole, as one piece, is inserted in its proper place by driving it down from the top on the inside. The whole process is finished by driving on the hoops, and making the holes in the handles.

9. The cedar employed in this business is a considerable tree, which grows in various parts of the world, but especially in the United States, where it occupies large tracts called *cedar* or *cypress swamps.* The wood is soft, smooth, and of an aromatic smell. It is likewise much used for shingles. The Dismal Swamp, lying in Virginia and North Carolina, contains an abundance of this kind of timber.

10. The operations in oak vary from those in cedar so far as to conform to the nature of the material, and the form of the vessels manufactured. In bringing the staves to the proper form, the workman is guided altogether by the eye; and, if they must be bent, they require to be heated. The fire for this purpose is made of shavings and chips in a small furnace of sheet iron, called a *crusset.* The hoops, both for ce-

dar and oak wares, are made of thin strips of iron, or of small oak, hickory, ash, or cedar saplings. Within a few years, several machines have been invented, for getting out staves, and for bringing them to the proper form, as well as for performing several other parts of the cooper's operations.

11. The coopers in England derive a great deal of their employment from the West India trade. Barrels, puncheons, and hogsheads, are carried out of the country filled with dry goods, and are returned filled with rum and sugar. In the United States, much work of this kind is done for the same market; but then the staves and heads are only fitted and marked here, to be afterwards put together in the West Indies.

THE WHEELWRIGHT.

1. The artisan who makes the wood-work of common wheel carriages, or the wheels of coaches, is denominated a wheelwright; but, under this head, we propose to include whatever we may say on constructing and finishing wheel carriages in general.

2. It must be evident, even to a superficial observer, that this business, in its different branches, occupies a large space in our domestic industry, since almost every farmer in the country owns a vehicle of some sort, and since the streets of our busy cities and towns exhibit, during a great part of the day, scenes of bustle occasioned, in a great measure, by the passing and repassing of carriages of different kinds.

3. The principal kinds of wheel carriages made in this country, are the cart, the wagon the gig, and

II.—N

the coach; and of each of these there are various sorts, differing in strength and mode of construction, to suit the particular purposes to which they are to be applied. The business of making these vehicles is divided into a number of branches; but, as the manufacture of the coach embraces a greater variety of operations than any other species of carriage, we have selected it as affording the best means of explaining the operations of the whole business.

4. In large establishments for making coaches and other vehicles of the best workmanship, the operators confine their attention to the execution of particular parts of the work; for example, one man makes the wheels, another the carriage and body, another fashions and applies the iron, another does the painting and polishing, and another the trimming. In smaller establishments, a greater proportion of the work is executed by one person.

5. The wheels of the coach, as well as those of every other vehicle in which they are used, are composed of a *hub*, and several *spokes*, and *felloes*. The hubs are commonly made of a kind of tough wood, called *gum*, which is reduced to the desired form in the lathe. The hole through the centre is made with a common auger, and enlarged with one tapering towards the point, and having through its whole length two cutting edges. The mortices for the spokes are made with a chisel driven with a mallet.

6. The spokes are made of white oak, and the felloes, of ash or hickory; and both are brought to the required form and smoothness with the saw, axe, drawing-knife, spoke-shave, chisel, and sand-paper. The constituent parts of the *carriage,* or *running gears,* are the *axles, perch,* and *spring-beds,* or *bolsters,* to which are added the *tongue,* or *pole,* and some other parts connected with it.

7. The joints in this part of the vehicle are made

perfectly tight by the application of putty; whereas, in the body, glue is used for this purpose. The latter substance will not answer in the former case, since it cannot bear exposure to water. The wood generally employed for the carriage part, as well as for the frame of the body, is ash; and the several parts are sawn from planks of suitable thickness. In this part of the work, the operator is guided by patterns made of thin pine boards. The panels of the body are made of thin boards of poplar or bass-wood. The manner in which the several parts are dressed and put together is too obvious to need description.

8. The wheels and the carriage, after having received one coat of paint, are sent to the blacksmith to be ironed. The hub is bound, at each end, with hoops of iron, commonly plated with brass or silver, and the outside rim or felloes are bound with an iron *tire*, and fastened with strong nails or spikes. The tires are made red-hot before they are applied, that they may be made to fit in every part with accuracy.

9. Bands, bolts, or strips of iron, are applied to those parts of the wood-work which may be exposed to friction, or which require additional strength. The axles are also made of wrought iron, either by the blacksmith who executes the other iron work, or by persons who manufacture them by the quantity for sale. The same remark is applicable to the *thorough-boxes*, which are inserted into the hub to prevent injury by friction, and to cause the wheel to revolve with freedom and accuracy.

10. The painting, varnishing, and polishing, of the body of the coach, when done in the best manner, comprise a tedious process. It is first covered with a coat of paint; the grain of the wood is then filled up with putty, and the surface is again covered with paint. Five coats of *filling*, composed of ochre, japan varnish, and spirits of turpentine, are next success-

ively applied. After the surface has been rubbed with a solid piece of pumice-stone, it is again painted, and rubbed with sand-paper. Several coats of paint are next laid on, and the work is finished by the application of a few coats of copal-varnish, and by the use of pumice-stone. The painting and varnishing of the wheels and carriage part, is far less expensive and tedious.

11. The nature of the trimmings, and the manner in which they are put together and applied, need not be described, since a few moments' inspection of a finished vehicle of this kind, will give any one a clear conception of the whole of this branch of the business. So far as trimming the inside, and the manufacture of cushions are concerned, the operations are similar to those of the upholsterer.

12. Wheel carriages may be classed among the primitive inventions, although the first authentic notice we have of their use, we find in the scripture history of Joseph, the son of Jacob, in which it is related, that this great and good man " was made to ride in the second chariot" of the king's, and that he sent wagons from Egypt to convey thither his father and family from the land of Canaan.

13. Covered wagons were used in the days of Moses; and the wandering Scythians, in the time of the Romans, had them covered with leather. The seat for the driver is said to have been invented by Oxylus, an Ætolian, who took possession of the kingdom of Elis, about 1100 years before Christ. Many of the nations of antiquity used chariots in the field of battle, and the axles were sometimes armed with scythes or some other sharp cutting instruments. Two persons commonly occupied one vehicle, one of whom drove the horses, and the other fought the enemy. The inhabitants of the promised land fought in chariots, even before the settlement of the people of Is-

racl in that country; and the Greeks likewise employed them, for warlike purposes, at the siege of Troy.

14. The carriages used by the Romans were of various kinds, some of which were carried on the shoulders of men, and others, having two or four wheels, were drawn by horses, asses, mules, or oxen. Nevertheless, neither they, nor any other nation of antiquity, ever suspended the body of any carriage on leathers, or supported it on springs; and the use of almost every species of vehicle for the conveyance of persons, was banished by the policy of the barbarous nations that afterwards became masters of civilized Europe, the feudal lords conceiving it important, that their military vassals should serve them on horseback.

15. Even as late as the sixteenth century, ministers rode to court, and magistrates of imperial cities to council, on the back of this animal; and, in the same manner, kings and lords made their public entry on the most solemn occasions. In accounts of papal ceremonies which occurred during several centuries, we find no mention of a state-coach; but, instead of it, state-horses or state-mules. The horse for his holiness was required to be a gentle and tractable nag, of a gray color; and a stool with three steps was necessary to aid him in mounting. The emperor or kings, if present, held his stirrup, and led his beast. Bishops also made their public entrance on horses or asses richly decorated.

16. Covered carriages, however, were known in the principal states of Europe in the fifteenth and sixteenth centuries; but they were at first used only by women of rank, since the men thought it disgraceful to ride in them. At this period, when the electors of the German empire did not choose to be present at the meetings of the states, they excused themselves to the emperor by stating that their health would not

permit them to ride on horseback, and it was not becoming for them to ride like women.

17. But, for a long time, the use of carriages was forbidden even to women; and, as late as the year 1545, the wife of a certain duke obtained from him, with great difficulty, the privilege of using a covered carriage in a journey to the baths. The permission was granted on the condition that her attendants should not enjoy the same favor. Nevertheless, it is certain that emperors, kings, and princes, began to employ covered carriages on journeys, in the fifteenth century; and a few instances occur of their use in public solemnities. Ambassadors appeared, for the first time, in coaches, at a public solemnity, in 1613, at Erfurth.

18. In the history of France, we find many proofs, that, in the fourteenth, fifteenth, and sixteenth centuries, the French monarchs commonly rode on horses, the servants of the court on mules, and the princesses, together with the principal ladies, sometimes at least, on asses. Carriages of some sort, however, appear to have been used at a very early period there. An ordinance of Philip the Fair, issued in 1294, forbids their use by the wives of citizens.

19. In the year 1550, three coaches were introduced into Paris; one of which belonged to the queen, another to Diana de Poictiers, and the third to Raimond de Laval, a cavalier of the court of Francis I., who was so large that no horse could carry him. It is not certain, however, that the body of these vehicles were suspended on leather straps. The inventor of this material improvement cannot be ascertained, nor is it positively determined, that it had been made, until about the middle of the seventeenth century.

20. Coaches were introduced into Spain and Portugal, in the year 1546, and into Sweden near the

close of the same century. In the capital of Russia, there were elegant coaches as early as the beginning of the seventeenth century. In Switzerland, they were rare, as late as 1650. Carriages began to be used at Naples in the thirteenth century; from this place they spread all over Italy; and here, also, glass panels originated.

21. Carriages of some sort were used in England at a very early period, and those first employed by the ladies, were called *whirlicoats*. According to some authors, coaches were introduced in the year 1555; but, according to others, not until twenty-five years after this period. Before the latter date, Queen Elizabeth, on public occasions, rode on the same horse with her chamberlain, seated behind him on a pillion; although, in the early part of her reign, she owned a chariot.

22. In 1601, men were forbidden the use of the coach by act of Parliament, the legislators supposing such indulgence to be too effeminate; but this law seems to have been little regarded, as this vehicle was in common use, about the year 1605. Twenty years after this time, hackney coaches began to ply in London; but these were prohibited, in 1635, on the alleged ground that the support of so many horses increased the expense of keeping those belonging to the king. Two years after this, however, fifty coaches were licensed, and, in 1770, there were one thousand.

23. The stage-coach was first employed in France, and was introduced into England, near the middle of the eighteenth century, by Jethro Tull, the celebrated agriculturist. They were not employed, in any country, in the transportation of the mail, until the year 1784. Before this time, it was carried chiefly on horseback.

24. In the United States, the manufacture of carriages of every kind has greatly increased within a few

years, and those lately made exhibit many improvements on those of former periods. The places which seem to be most distinguished for the manufacture of good carriages, in this country, are Philadelphia, Newark, and Troy.

THE POTTER.

1. THE artisan called the potter converts plastic materials into hard and brittle vessels of various kinds, denominated, in general terms, *earthen ware.*

2. Alumine is the basis of all clays, and is the only earth that possesses the degree of plasticity which renders the operations of the potter practicable. It is, however, never found or used in a pure state, but in combination with other substances, particularly with silex, lime, magnesia, and the oxyde of iron.

3. In the manufacture of vessels from argillaceous compounds, the different degrees of beauty and costliness depend upon the quality of the raw materials, and the labor and skill expended in the operation. The various productions of the pottery may be classed under the following denominations—common earth-

en ware, white earthen ware, stone ware, and porcelain; but of each of these there are many varieties.

4. *Common earthen ware.*—This ware is made of a kind of clay very generally diffused over the earth, and which is essentially the same with that employed in making bricks. The potters are often supplied with this material by the brickmakers, who select for them that which is too tenacious, or *fat*, for their own purpose. All common clays contain more or less of the oxyde of iron, which causes the wares made of them to turn red in burning.

5. In preparing the clay for use, the potter adds to it, when necessary, a portion of fine loam, in order to lessen its tenacity, and to prevent the vessels to be made of it from cracking, while undergoing the fire. When the materials have been mixed, and partially incorporated with water, the mass is thrown into a tub, fixed in the ground about one-half of its depth. In the centre of this tub, is placed a shaft, in a perpendicular position, from which radiate, in a horizontal direction, a number of knives or cutters.

6. This machine is put in motion by horse-power, and by it the clay is repeatedly cut, and properly kneaded. The workman then cuts it into thin slices with a small wire, and, having rejected all matters not fit for his purpose, he further kneads it with his hands, and forms it into lumps, corresponding in amount of matter with the different vessels which he proposes to make.

7. For the best kinds of this ware, the same species of clay is used; but then it is differently prepared. It is first dissolved in water; and, when the coarser particles have settled to the bottom of the vessel, the fluid suspending the rest is drawn off, and made to pass through a sieve into a reservoir. After the particles of the material have precipitated, the water is drawn off, and the residuum is thrown upon

a large flat pan or reservoir made of bricks, where the mass is freed from its superfluous moisture by evaporation in the air, or by means of artificial heat applied beneath. It is then laid by in a damp place, for future use.

8. Before the clay, thus purified from extraneous and coarser particles, is formed into vessels, it is beaten with a stout piece of wood, until the mass has become of an equal consistence throughout, and then repeatedly cut into two pieces with a wire, and slapped together to expel the air. The former of these operations is called *wedging*, and the latter, *slapping*.

9. *White and cream-colored wares* are made of clays which contain so little oxyde of iron, that it does not turn red in burning, but, on the contrary, improves in whiteness in the furnace. There are several species of white clay, found in many different localities, most of which, however, are known under the denomination of *pipe-clay;* or they are distinguished by the names of the places where they are obtained.

10. In preparing these clays for use, they are reduced to a minute division by machinery, and afterwards dissolved in water, and otherwise treated in a manner similar to that used for the better kinds of common wares, as described in the seventh and eighth paragraphs. For the purpose of diminishing the shrinkage in the fire, and with the view of increasing the whiteness of the ware, pulverized flint-stone is added to the clay, in the proportion of about one part of the former to five of the latter.

11. In reducing the silex to the requisite fineness, it is first brought to a red heat; and, while in this state, it is thrown into cold water, to diminish the cohesion of its parts. It is then pounded by machinery, levigated with water in a mill, sifted, mashed, and

otherwise treated like the clay. The materials are mixed while in a state of thin pulp.

12. The several operations performed by the potter, in converting the clay thus prepared into different kinds of vessels, and in completing the whole process of the manufacture of earthen ware, may be included under the following divisions, viz., throwing, turning, pressing, burning, painting and printing, and glazing. They are not, however, all used in producing and finishing vessels of every shape and quality.

13. *Throwing.*—This operation is performed on a potter's wheel, which consists of a round table, and some simple means to put it in motion. The clay having been placed on the centre of this machine, the workman communicates to the latter a rotary motion with his foot, and gives the proposed form to the material with his hands, which have been previously wet with water, to prevent them from sticking. This method is used for all vessels and parts of vessels of a circular form; and, in many cases, no other operation is necessary to give them the requisite finish, so far as their conformation is concerned.

14. *Turning.*—The vessels are cut from the thrower's wheel with a small wire; and when, by the evaporation of moisture, they have become firm enough to endure the operation, they are turned on a lathe. The objects of this operation are to communicate to them a more exact shape, and to render them more uniform in thickness. The potter's wheel, with the addition of some contrivance to hold the pieces in a proper position, is frequently used for turning. The coarser kinds of common wares are never turned.

15. *Pressing.*—Vessels, or parts of vessels, which are of an irregular shape, and which cannot be formed on the wheel, are usually made by a process called *pressing*. This kind of work is executed in moulds made of plaster of Paris, and these are formed on

models of clay or wood, which have been made in the exact shape of the proposed vessel. Sometimes individual specimens of the wares of one country or pottery are used as models in another; in such cases, the expense of the moulds is considerably diminished.

16. The moulds frequently consist of several parts, which fit accurately together; for example, the mould for a pitcher is composed of two pieces for the sides, and one for the bottom. In forming a pitcher in such a mould, the material, which has been spread out to a proper and uniform thickness, is laid upon the inside of each portion of it, and the superfluous clay is trimmed off with a knife. The mould is then closed, and thin strips of clay are laid over the seams; the removal of the several pieces of the mould, completes the operation.

17. Handles, spouts, figures in relief, and other additions of this nature, are separately made in moulds, and stuck on the vessel with the same kind of materials, sometimes mingled with a small proportion of plaster of Paris. These appendages are added after the vessels have become partially solid in the air.

18. *Burning.*—All vessels, even after they have been dried in the atmosphere, are in a very frangible state; and, to render them sufficiently firm for use, they are submitted to the process of burning in a kiln. To preserve the ware from injury while enduring the fire, the several pieces are enclosed in cylindrical boxes called *saggers*, which are made of baked clay. These boxes are placed one above another around the sides of the kiln, which is of a circular form, and gradually tapering to the top.

19. In burning the coarser wares, every piece is not thus inclosed; but, between every two saggers, a naked piece is placed. A moderate fire is first raised,

which is gradually increased, until the contents of the kiln are brought to a red heat. The burning occupies between twenty-four and forty-eight hours. All wares, except the coarsest kinds, are twice, and sometimes thrice, burned; and, after having been once submitted to the process, they are said to be in a state of *biscuit.*

20. *Painting and printing.*—When the vessels are to be ornamented with colors, it is necessary, in most cases, that this part of the work be done after the first burning. In China, and at the porcelain manufactory in Philadelphia, the drawings are executed by hand with a pencil. The same method is used in Europe in elaborate pieces of workmanship. But, in the common figured wares, where but one color is used, the designs are first engraved on metallic plates, and impressions are taken from them on thin paper, by means of a copperplate printing-press.

21. In transferring to vessels designs thus produced, the paper, while in a damp state, is applied closely to the surface of the biscuit, and rubbed on with a piece of flannel. The porosity of the earthen material causes the immediate absorption of the coloring matter, which, in all cases, is some metallic oxyde. For a blue color, the oxyde of cobalt is used; and for a black, those of manganese and iron. The paper is washed from the ware with a sponge.

22. *Glazing.*—To prevent the penetration of fluids, and to improve the appearance of the ware, a superficial vitreous coating is necessary. This can be produced by the aid of various substances; but, in a majority of cases, red lead is the basis of the mixture employed for this purpose. Equal parts of ground flints and red lead are used for the common cream-colored wares. These materials are mixed with, and suspended in, water, and each piece is dipped in the liquid. The moisture is soon absorbed by the clay

leaving the glazing particles on the surface, which, in the burning that follows, is converted into a uniform and durable vitreous coating.

23. *Stone ware.*—The materials of this ware, as well as the mode of preparing them, differ but little from those of the common and better kinds of earthen wares. The clays, however, which contain but little or no oxyde of iron are chosen, since this substance would cause the ware to melt and warp, before a sufficient degree of heat could be applied to give it the requisite hardness.

24. The glazing is formed by a vitrification of the surface of the vessels, caused by the action of common salt thrown into the kiln, when it has been raised to its greatest heat. This glazing is more perfect than that on ordinary earthen wares, being insoluble by most chemical agents. It is hardly necessary to remark that this method of glazing precludes the use of saggers.

25. *Porcelain.*—This ware exceeds every other kind in the delicacy of its texture, and is peculiarly distinguished by a beautiful semi-transparency, which is conspicuous when held against the light. In China, it is made chiefly of two kinds of earth; one of which is denominated *petuntze,* and the other *kaolin;* but both are varieties of feldspar, found in the mountains, in different localities. They are brought to the manufactories from a distance in the form of bricks; the materials, as taken from the mines, having been reduced to an impalpable powder in mortars, either by the labor of men or by water-power.

26. These materials are combined in different proportions in the manufacture, according to the quality of the proposed ware. In the best kind, equal quantities are used; but for those of inferior quality, a greater proportion of petuntze is employed. The translucency so much admired in porcelain, or *tseki,*

as the Chinese call it, is owing to the petuntze, which, in burning, partially melts, and envelops the infusible kaolin.

27. It is not known who was the inventor of porcelain, as the Chinese annals are silent with regard to this point; nor do we know more of the date at which the manufacture was commenced. It is certain, however, that it must have been before the fifth century of the Christian era. Since this ware has been known to Europeans, it has been manufactured chiefly, and in the greatest perfection, in the large and populous village of King-te-ching.

28. Porcelain was first brought to Europe from Japan and China, and for a long time its materials and mode of manufacture remained a secret, in spite of the efforts of the Jesuit missionaries, who resided in those countries. At length, in 1712, Father Entrecolles sent home to France, specimens of petuntze and kaolin, together with a summary description of the process of the manufacture.

29. Shortly after this important event had transpired, it was discovered that materials nearly of the same kind existed in abundance in various parts of Europe. The manufacture of porcelain was, therefore, soon commenced in several places; and it has since been successfully carried on.

30. The porcelain wares of Europe are superior to those of the Chinese, in the variety and elegance of their forms, as well as in the beauty of the designs executed upon them; but, as some of the processes successfully practised in China, remain still to be learned by the Europeans, the Oriental porcelain has not yet been equalled in the hardness, strength, and durability of its body, and in the permanency of its glaze. The manufacturers of Saxony are said to have been the most successful in their imitations in these respects.

31. The porcelain earths are found in various parts

of the United States, but particularly at Wilmington, in the state of Delaware. Nevertheless, there is now but one porcelain manufactory in our country, and this is yet in its infancy. The establishment is located in Philadelphia, and it has been lately incorporated, with the privilege of one hundred thousand dollars capital.

32. The principle of induration by heat, is the same in the manufacture of earthen wares as in making bricks; and, as the latter can be more easily dispensed with than the former in a primitive state of society, it is but reasonable to suppose that earthen ware was first invented; but the art of making bricks must have been practised before the deluge, or the posterity of Noah would not have attempted so soon as about one hundred years after that catastrophe, to build a city and a tower of these materials. It is, therefore, evident, that this art was of antediluvian origin; and it was probably one of the earliest brought to any degree of perfection.

33. The art of the potter was practised more or less by every nation of antiquity, and the degree of perfection to which it was carried in every country corresponded with the state of the arts generally. The Greeks were consequently very celebrated for their earthen wares. The Etruscans have also been particularly noted for their manufacture of the elegant vases which have been dug, in modern times, from the depositories of the dead, in Lower Italy.

34. Until the commencement of the manufacture of porcelain in Europe, this art continued in a very rude condition, although practised to a considerable extent in many places. It was much improved in England about the year 1720, by the addition of flints to the usual material; and, between thirty and forty years after this, it was brought to great perfection, in all its branches, chiefly through the scientific exertions of the celebrated potter, Josiah Wedgewood.

THE GLASS-BLOWER.

1. GLASS is a substance produced from a combination of silicious earths with alkalies, and, in many cases, with metallic oxydes. The basis of every species of glass is silex, which is found in a state nearly pure in the sands of many situations. It is also found in the common flints and quartz pebbles.

2. When quartz pebbles or flints are employed, they must be first reduced to powder. This is done by grinding them in a mill, after they have been partially reduced, by heating them in the fire, and plunging them into cold water. Sand has the advantage of being already in a state of division sufficiently minute for the purpose. To prepare it for application, it only requires to be washed and sifted, in order to free it from the argillaceous and other substances unfit for

use. A great proportion of the sand employed in the manufacture of the better kinds of glass in the United States, is taken from the banks of the Delaware River.

3. The alkaline substances used are potash and soda. For the finer kinds of glass, pearlash, or soda procured by decomposing sea-salt, is used; but, for the inferior sorts, impure alkalies, such as barilla, Scotch and Irish kelp, and even wood-ashes, as well as the refuse of the soap-boiler's kettle, are made to answer the purpose. Lime, borax, and common salt, are also frequently used as a flux in aid of some of the other substances just mentioned.

4. Of the metallic oxydes which make a part of the materials of some glass, the deutoxyde of lead, or, as it is usually denominated, red lead, is the most common. This substance is employed in making flint glass, which is rendered by it more fusible, heavy and tough, and more easy to be ground or cut, while, at the same time, it increases its brilliancy and refractive power.

5. Black oxyde of manganese is also used in small quantities, with the view of rendering the glass more colorless and transparent. Common nitre produces the same effect. White arsenic is also added to the materials of this kind of glass, to promote its clearness; but, if too much is used, it communicates a milky whiteness. The use of this substance in drinking vessels is not free from danger, when the glass contains so much alkali as to render any part of it soluble in acids.

6. The furnace in which the materials are melted is a large conical stack, such as is represented at the head of this article. In some cases, it is surrounded by a large chimney, which extends above the roof of the building. In the sides are several apertures, near which are placed the crucibles, or melting-pots, containing the materials. The fuel is applied in an arch,

which is considerably lower than the surface of the ground on which the operators stand, while at work.

7. The melting-pots are made chiefly of the most refractory clays and sand. Much of the clay used for this purpose, in many of the glass-houses in the United States, is imported from Germany. The materials, having been sifted, and mixed with a suitable quantity of water, the homogeneous mass is formed into crucibles, by spreading it on the inside of vessels which are much in the shape of a common wash-tub. After the clay has become sufficiently solid to sustain itself, the hoops are removed from the vessel, and the several staves taken apart.

8. The crucibles are suffered to dry in the atmosphere for two or three months, after which they are applied to use as they may be needed. Before they are placed in the main furnace, they are gradually raised to an intense heat in one of smaller dimensions, built for this express purpose. The fuel employed in fusing the *metal* is chiefly pine wood, which, in all cases, is previously dried in a large oven. Four of the five furnaces near Philadelphia, which belonged to Doctor Dyott, were heated with rosin.

9. The materials having been mixed, in the proposed proportions, which are determined by weight, they are thrown into the melting-pots, and, by a gradually increasing heat, reduced to a paste, suitable for application by the blower. This part of the process is commonly performed at night, while the blowers are absent from the works.

10. The applications of glass are so exceedingly extensive, that it is inconvenient, if not impossible, to manufacture every species of it at one glass-house or at one establishment. Some, therefore, confine their attention to the production of window glass, and such articles of hollow ware as may be made, with profit, from the same kind of paste. Others make

vials and other species of ware, employed by the druggist, apothecary, and chemist. And again, the efforts, at some factories, are confined entirely to the manufacture of flint glass, or to that of plate glass for mirrors.

11. The principal operations connected with the manufacture of different species of glass, after the paste has been prepared, may be included under the following heads; viz., blowing, casting, moulding, pressing and grinding; although all these are never performed in one and the same establishment.

12. *Blowing.*—The operation of blowing is nearly or quite the same in the production of every species of glass ware, in which it is employed. The manipulations, however, connected with making different articles, are considerably varied, to suit their particular conformation. This circumstance renders it impossible for us to give more than a general outline of the process of this manufacture.

13. In the formation of window glass, the workman gathers upon the end of an iron tube a sufficient amount of the metal, which he brings to a cylindrical form by rolling it upon a cast iron or stone table. He then blows through the tube with considerable force, and thus expands the glass to the form of an inflated bladder. The inflation is assisted by the heat, which causes the air and moisture of the breath to expand with great power.

14. Whenever the glass has become too stiff, by cooling, for inflation, it is again softened by holding it in the blaze of the fuel, and the blowing is repeated, until the globe has been expanded to the requisite thinness. Another workman next receives it at the other end, upon an iron rod, called a *punt*, or *punting iron*, when the blowing iron is detached. It is now opened, and spread into a smooth sheet, by the centrifugal force acquired by the rapid whirl given to it, in the manner exhibited in the preceding cut. The sheet

thus produced is of a uniform thickness, except at the centre, where the iron rod had been attached.

15. An inferior kind of window glass, the materials of which are sand, kelp, and soap-boilers' waste, is made by blowing the *metal* into cones, about a foot in diameter at their base; and these, while hot, are touched on one side with a cold iron dipped in water. This produces a crack, which runs through the whole length of the cone. The glass then expands into a sheet somewhat resembling a fan. This is supposed to be the oldest method of manufacturing window or plate glass.

16. The window glass produced in the manner first described, is called *crown glass;* and the other, *broad glass.* But by neither of these methods can the largest panes be produced. The blowing for these differs from the methods just described, in that the material is blown into an irregular cylinder, open at its further end. When a sufficient number of these cylinders have accumulated, the end to which the blowing iron had been attached, is *capped off* by drawing round it a circle of melted glass, and the cylinder is divided longitudinally by touching it through its whole length with a hot iron. The cylinders, in this state, are put into the annealing oven, where, by aid of a heat which raises the glass to redness, it is expanded into sheets. These sheets are then broken into panes of several sizes by the aid of a diamond and a straight edge, as in the case of glass blown by other methods.

17. *Casting.*—Plate glass formed by the method last mentioned, is denominated *cylinder glass;* and it is used not only for windows, but also for mirrors not exceeding four feet in length. Plates of greater dimensions are produced by a process called *casting.* The casting is performed by pouring the material, in a high state of fusion, upon a table of polished copper of large size, and having a rim elevated above its gen-

eral surface, as high as the proposed plate is to be thick. To spread the glass perfectly, and to render the two surfaces parallel, a heavy roller of polished copper, resting upon the rim at the edges, is passed over it.

18. Plates thus cast are always dull and uneven. To render them good reflectors, it is necessary to grind and polish them. The plate to be polished is first cemented with plaster of Paris to a table of wood or stone. A quantity of wet sand, emery, or pulverized flints, is spread upon it, and another glass plate, similarly cemented to a wooden or stone surface, is placed upon it. The two plates are then rubbed together, until their surfaces have become plane and smooth. The last polish is given by colcothar and putty. Both sides are polished in the same manner.

19. *Moulding.*—Ornamental forms and letters are produced on the external surface of vessels, by means of metallic moulds; and the process by which this kind of work is performed is called moulding. In the execution, the workman gathers upon the end of his iron tube, a proper amount of the material, which he extends, and brings to a cylindrical form, by rolling it upon his table. He then expands it a little by a slight blast, and afterwards lets it down into the mould, which is immediately filled by blowing still stronger through the tube.

20. The vessel is then taken from the mould, and disengaged from the tube. The same tube, or a punting iron having been attached to the bottom, the other end is softened in the fire, and brought to the proposed form with appropriate tools, while the iron is rolled up and down upon the long arms of the glass-blower's chair. The ornamental moulds are made of cast iron, brass, or copper, and are composed of two parts, which open and shut upon hinges. The moulds for plain vials, castor oil bottles, small demijohns, &c., are

made of the kind of clay used for the crucibles These consist merely of a mass of the clay, with a cylindrical hole in it of proper diameter and depth.

21. *Pressing.*—This process is applied in the production of vessels or articles which are very thick, and which are not contracted at the top. The operation is performed in iron moulds, which consist of two parts, and which have upon their internal surfaces the figures to be impressed upon the glass. The material, while in an elastic condition, is put into the lower part of the mould; and the other part, called the *follower*, is immediately brought upon it with considerable force.

22. Every species of glass, before it can be used with safety, must be *annealed*, to diminish its brittleness. The annealing consists merely in letting down the temperature by degrees. Small boys, therefore, convey the articles, whatever they may be, as fast as they are made, to a moderately heated oven, which, when filled, is suffered to cool by degrees.

23. *Cutting.*—The name of *cut glass* is given to the kind which is ground and polished in figures, appearing as if cut with a sharp instrument. This operation is confined chiefly to flint glass, which, being more tough and soft than the other kinds, is more easily wrought. In addition to this, it is considerably more brilliant, producing specimens of greater lustre.

24. An establishment for grinding glass contains a great number of wheels of cast iron, stone, and wood, of different sizes; and the process consists entirely in holding the glass against these, while they are revolving with rapidity. When a considerable portion of the material is to be removed, the grinding is commonly commenced on the iron wheel, on which is constantly pouring water and sharp sand, from a vessel above, which, from its shape, is called a *hopper*.

25. The period of the invention of glass is quite unknown; but the following is the usual story of its

origin. Some merchants, driven by a storm upon the coasts of Phœnicia, near the River Belus, kindled a fire on the sand to cook their victuals, using as fuel some weeds which grew near. The ashes produced by the incineration of these plants, coming in contact with the sand, united with its particles, and, by the influence of the heat, produced glass.

26. This production was accidentally picked up by a Tyrian merchant, who, from its beauty and probable utility, was led to investigate the causes of its formation, and who, after many attempts, succeeded in the manufacture of glass. The legend probably originated in the fact, that glass was very anciently made at Tyre; and that the sand on the seashore in the immediate neighbourhood of the Belus, was well adapted to glass-making.

27. It is certainly probable, that an accidental vitrification might have given rise to the discovery; but the circumstance would have been more likely to take place in some operation requiring greater heat than that necessary for dressing food in the open air. The invention of glass must have been effected as early as fifteen hundred years before our era. It was manufactured very anciently in Egypt; but whether that country or Phœnicia is entitled to the preference, as regards priority in the practice of this art, cannot be determined.

28. Glass was made in considerable perfection at Alexandria, and was thence supplied to the Romans as late as the first quarter of the second century. Before this time, however, Rome had her glass manufactories, to which a particular street was assigned. The attention of the workmen was directed chiefly to the production of bottles and ornamental vases, specimens of which still remain, as monuments of their extraordinary skill.

29. In modern times, the manufacture of glass was

confined principally to Italy and Germany. Venice became particularly celebrated for the beauty of the material, and the skill of its workmen; and as early as the thirteenth century, it supplied the greatest part of the glass used in Europe. The artists of Bohemia, also, came to be held in considerable reputation.

30. The art was first practised in England, in the year 1557, when a manufactory was erected at Crutched Friars, in the city of London, and shortly afterwards, another at the Savoy, in the Strand. In these establishments, however, were made little else than common window glass, and coarse bottles, all the finer articles being still imported from Venice. In 1673, the celebrated Duke of Buckingham brought workmen from Italy, and established a manufactory for casting plate glass for mirrors and coach windows. The art, in all its branches, is now extensively practised in great perfection, not only in Great Britain, but in many of the other kingdoms of Europe.

31. Before the commencement of the late war with England, very little, if any, glass was manufactured in the United States, except the most common window glass, and the most ordinary kinds of hollow ware. Apothecaries' vials and bottles, as well as every other variety of the better kinds of glass wares, had been imported from Europe, and chiefly from England.

32. Our necessities, created by the event just mentioned, produced several manufactories, which, however, did not soon become flourishing, owing, at first, to inexperience, and, after the peace, to excessive importations. But adequate protection having been extended to this branch of our national industry, by the tariff of 1828, it is now in a highly prosperous condition—so much so, that importations of glass ware have nearly ceased.

THE OPTICIAN.

1. THE word optician is applicable to persons who are particularly skilled in the science of vision, but especially to those who devote their attention to the manufacture of optical instruments, such as the spectacles, the camera obscura, the magic lantern, the telescope, the microscope, and the quadrant.

2. Light is an emanation from the sun and other luminous bodies, and is that substance which renders opaque bodies visible to the eye. It diverges in a direct line, unless interrupted by some obstacle, and its motion has been estimated at *two hundred thousand miles* in a second.

3. A *ray of light* is the motion of a single particle: and a parcel of rays passing from a single point, is called *a pencil of rays*. *Parallel rays* are such as always move at the same distance from each other.

Rays which continually approach each other, are said to *converge*; and when they continually recede from each other, they are said to *diverge*. The point at which converging rays meet is called the *focus*.

4. Any pellucid or transparent body, as air, water, and glass, which admits the free passage of light, is called a *medium*. When rays, after having passed through one medium, are bent out of their original course by entering another of different density, they are said to be *refracted*; and when they strike against a surface, and are sent back from it, they are said to be *reflected*.

5. A *lens* is glass ground in such a form as to collect or disperse the rays of light which pass through it. These are of different shapes; and they have, therefore, received different appellations. A *plano-convex* lens has one side flat, and the other convex; a *plano-concave* lens is flat on one side, and concave on the other; a *double convex* lens is convex on both sides; a *double concave* lens is concave on both sides; a *meniscus* is convex on one side, and concave on the other. By the following cut, the lenses are exhibited in the order in which they have been mentioned.

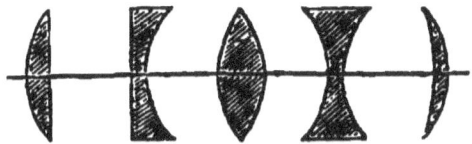

6. An *incident ray* is that which comes from any luminous body to a reflecting surface; and that which is sent back from a reflecting surface, is called a *reflected ray*. The *angle of incidence* is the angle which is formed by the incident ray with a perpendicular to the reflecting surface; and the *angle of reflection* is the angle formed by the same perpendicular and the reflected ray.

7. When the light proceeding from every point of an object placed before a lens is collected in corresponding points behind it, a perfect image of the object is there produced. The following cut is given by way of illustration.

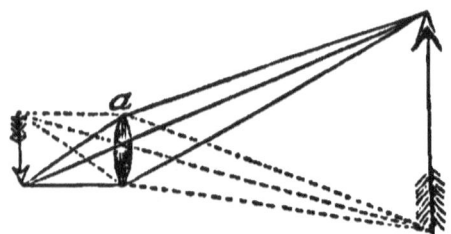

8. The lens, *a*, may be supposed to be placed in the hole of a window-shutter of a darkened room, and the arrow at the right to be standing at some distance without. All the light reflected from the latter object towards the lens, passes through it, and concentrates, within the room, in a focal point, at which, if a sheet of paper, or any other plane of a similar color, is placed, the image of the object will be seen upon it.

9. This phenomenon is called the *camera obscura*, or dark chamber, because it is necessary to darken the room to exhibit it. The image at the focal point within the room is in an inverted position. The reason why it is thrown in this manner will be readily understood by observing the direction of the reflected rays, as they pass from the object through the lens. In the camera obscura, it is customary to place a small mirror immediately behind the lens, so as to throw all the light which enters, downwards upon a whitened table, where the picture may be conveniently contemplated.

10. From the preceding explanation of the camera obscura, the theory of vision may be readily comprehended, since the eye itself is a perfect instrument of this kind. A careful examination of the following

representation of the eye will render the similarity obvious. The eye is supposed to be cut through the middle, from above downwards.

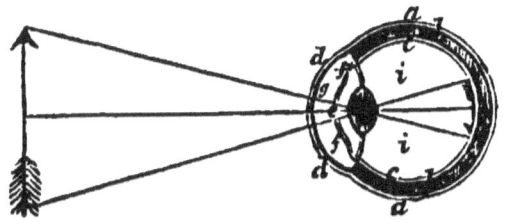

a a, the *sclerotica*; *b b*, the *choroides*; *c c*, the *retina*; *d d*, the *cornea*; *e*, the *pupil*; *f f*, the *iris*; *g*, the *aqueous humor*; *h*, the *crystalline humor*; *i i*, the *vitreous humor*.

11. The *sclerotica* is a membranous coat, to which the muscles are attached which move the eye. The *cornea* is united to the sclerotica around the circular opening of the latter, and is that convex part of the eye, which projects in advance of the rest of the organ. The space between this and the crystalline lens is occupied by the aqueous humor and the iris. The *iris* is united to the choroides, and it possesses the power of expanding and contracting, to admit a greater or less number of rays.

12. The *crystalline lens* is a small body of a crystalline appearance and lenticular shape, whence its name. It is situated between the aqueous and vitreous humors, and consists of a membranous sack filled with a humor of a crystalline appearance. The *vitreous humor* has been thus denominated on account of its resemblance to glass in a state of fusion. The *retina* is a membrane which lines the whole cavity of the eye, and is formed chiefly, if not entirely, by the expansion of the optic nerve.

13. The rays of light which proceed from objects pass through the cornea, aqueous humor, crystalline lens, and vitreous humor, and fall upon the retina in a focal point, to which it is brought, chiefly by the in-

fluence of the cornea and the crystalline lens. The image, in an inverted position, is painted or thrown on the cornea, which perceives its presence, and conveys an impression of it to the brain, by means of the optic nerve.

14. *Optical instruments.*—The art of constructing optical instruments is founded upon the anatomical structure, and physiological action of the eye, and on the laws of light. They are designed to increase the powers of the eye, or to remedy some defect in its structure. In the cursory view which we may give of a few of the many optical instruments which have been invented, we will begin with the *spectacles*, since they are the best known, and withal the most simple.

15. The *visual point*, or the distance at which small objects can be distinctly seen, varies in different individuals. As an average, it may be assumed at eight or nine inches from the eye. In some persons, it is much nearer, and in others, considerably more distant. The extreme, in the former case, constitutes *myopy*, or *short-sightedness*, and, in the latter case, *presbyopy*, or *long-sightedness*.

16. *Myopy* is chiefly caused by too great a convexity of the cornea and the crystalline lens, which causes the rays to converge to a focus, before they reach the retina. Objects are, therefore, indistinctly seen by myoptic persons, unless held very near the eye to throw the focus farther back. This defect may be palliated by the use of concave glasses, which render the rays proceeding from objects more divergent.

17. *Presbyopy* is principally caused by too little convexity of the cornea and crystalline lens, which throws the focal point of rays reflected from near objects, beyond the retina. This defect is experienced by most people, to a greater or less degree, after they have advanced beyond the fortieth year, and occa-

sionally even by youth. A remedy, or, at least, a palliation, is found in the use of convex glasses, which render the rays more convergent, and enable the eye to refract them to a focus farther forward, at the proper point.

18. The opticians have their spectacles numbered, to suit different periods of life; but, as the shortsighted and long-sighted conditions exist in a thousand different degrees, each person should select for himself such as will enable him to read without effort at the usual distance.

19. The great obstacle to viewing small objects at the usual distance, arises from too great a divergence of the light reflected from them, which causes the rays to reach the retina before they have converged to a focus. This defect is remedied by convex lenses, which bring the visual point nearer to the eye, and consequently cause the rays to concentrate in a large focus upon the retina. The most powerful microscopic lenses are small globules of glass, which permit the eye to be brought very near to the object.

20. *Microscopes* are either *single* or *double*. In the former case, but one lens is used, and through this the object is viewed directly; but, in the latter case, two or more glasses are employed, through one of which a magnified image is thrown upon a reflecting surface, and this is viewed through the other glass, or glasses, as the real object is seen through a single microscope.

21. The *solar microscope*, on account of its great magnifying powers, is the most wonderful instrument of this kind. The principles of its construction are the same with those of the camera obscura. The difference consists chiefly in the minor circumstance of placing the object very near the lens, by which a magnified image is thrown at the focal point within the room.

22. In the case of the camera obscura, the objects are at a far greater distance from the glass on the outside than the images, at the focal point, on the inside. The comparatively great distance of the object, in this case, causes the image to be proportionably smaller. In the solar microscope, a small mirror is used to receive the rays, and to reflect them directly upon the object.

23. The *magic lantern* is an instrument used for magnifying paintings on glass, and for throwing their images upon a white surface in a darkened room. Its general construction is the same with that of the solar microscope; but, in the application, the light of a lamp is employed instead of that from the sun.

24. *Telescopes* are employed for viewing objects which from their distances appear small, or are invisible to the naked eye. They are of two kinds, *refracting* and *reflecting*. The former kind is a compound of the camera obscura and the single microscope. It consists of a tube, having at the further end a double convex lens, which concentrates the rays at a focal point within, where the image is viewed through a microscopic lens, placed at the other end.

25. In the construction of reflecting telescopes, concave mirrors, or specula, are combined with a double convex lens. A large mirror of this kind is so placed in the tube, that it receives the rays of light from objects, and reflects them upon another of a smaller size. From this they are thrown to a focal point, where the image is viewed through a double convex lens. The specula are made of speculum metal, which is a composition of certain proportions of copper and tin.

26. Many optical appearances are of such frequent recurrence, that they could not have escaped the earliest observers; nevertheless, ages appear to have elapsed, before any progress was made towards an

explanation of them. Empedocles, a Greek philosopher, born at Agrigentum in Sicily, 460 years before Christ, is the first person on record who attempted to write systematically on light.

27. The subject was successively treated by several other philosophers; but the ancients never attained to a high degree of information upon it. We have reason to believe, however, that convex lenses were, in some cases, used as magnifiers, and as burning glasses, although the theory of their refractive power was not understood.

28. The magnifying power of glasses, and some other optical phenomena, were largely treated by Al Hazen, an Arabian philosopher, who flourished about the year 1100 of our era; and, in 1270, Vitellio, a Polander, published a treatise on optics, containing all that was valuable in Al Hazen's work, digested in a better manner, and with more lucid explanations of various phenomena.

29. Roger Bacon, an English monk, who was born in 1214, and who lived to the age of seventy-eight, described very accurately the effects of convex and concave lenses, and demonstrated, by actual experiment, that a small segment of a glass globe would greatly assist the sight of old persons. Concerning the actual inventor of spectacles, however, we have no certain information; we only know that these useful instruments were generally known in Europe, about the beginning of the fourteenth century.

30. In the year 1575, Maurolicus, a teacher of mathematics, at Messina, published a treatise on optics, in which he demonstrated that the crystalline humor of the eye is a lens, which collects the rays of light from external objects, and throws them upon the retina. Having arrived at a knowledge of these facts, he was enabled to assign the reasons why some people were short-sighted, and others long-sighted.

31. John Baptista Porta, of Naples, was contemporary with Maurolicus. He invented the camera obscura, and his experiments with this instrument convinced him, that light was a substance, and that its reception into the eye produced vision. These discoveries corresponded very nearly with those by Maurolicus, although neither of these philosophers had any knowledge of what the other had done. The importance of Porta's discoveries will be evident, when it is observed, that, before his time, vision was supposed to be dependent on what were termed *visual rays*, proceeding from the eye.

32. The telescope was invented towards the latter end of the sixteenth century. Of this, as of many other valuable inventions, accident furnished the first hint. It is said, that the children of Zacharias Jansen, a spectacle-maker, of Middleburg in Holland, while playing with spectacle-glasses in their father's shop, perceived that, when the glasses were held at a certain distance from each other, the dial of the clock appeared greatly magnified, but in an inverted position.

33. This incident suggested to their father the idea of adjusting two of these glasses on a board, so as to move them at pleasure. Two such glasses inclosed in a tube completed the invention of the simplest kind of the refracting telescope. Galileo greatly improved the telescope, and constructed one that magnified thirty-three times, and with this he made the astronomical discoveries which have immortalized his name.

34. John Kepler, a great mathematician and astronomer, who was born at Weir, in Wurtemburg, in the year 1571, paid great attention to the phenomena of light and vision. He was the first who demonstrated that the degree of refraction suffered by light in passing through lenses, corresponds with the diameter of the circle of which the concavity or convexity is the

portion of an arch. He very successfully pursued the discoveries of Maurolicus and Porta, and asserted that the images of external objects were formed upon the optic nerve by the concentration of rays which proceed from them.

35. In 1625, the curious discovery of Scheiner was published, at Rome, which placed beyond doubt the fact, that vision depends upon the formation of the image of objects upon the retina. The fact was demonstrated by cutting away, at the back part, the two outside coats of the eye of an animal, and by presenting different objects before it. The images were distinctly seen painted on the naked retina.

36. Near the middle of the seventeenth century, the velocity of light was discovered by Roemer; and, in 1663, James Gregory, a celebrated Scotch mathematician, published the first proposal for a reflecting telescope. But, as he possessed no mechanical dexterity himself, and as he could find no workman capable of executing his designs, he never succeeded in carrying his conceptions into effect. This was reserved for Sir Isaac Newton; who, being remarkable for manual skill, executed two instruments of this kind, in the year 1672, on a plan, however, somewhat different from that proposed by Gregory.

37. In the course of the year 1666, the attention of Sir Isaac Newton was drawn to the phenomena of the refraction of light through the prism; and, having observed a certain surprising fact, he instituted a variety of experiments, by which he was brought to the conclusion, that light was not a homogeneous substance, but that it is composed of particles, which are capable of different degrees of refrangibility.

38. By the same experiments, he also proved, that the rays or particles of light differ from each other in exhibiting different colors, some producing the color red, others that of yellow, blue, &c. He applied his

principles to the explanation of most of the phenomena of nature, where light and color are concerned; and almost every thing which we know upon these subjects, was laid open by his experiments.

39. The splendor of Sir Isaac Newton's discoveries obscures, in some measure, the merits of earlier and subsequent philosophers; yet several interesting discoveries in regard to light and color, as well as many important improvements of optical instruments, have been made since his time, although the light by which these have been achieved, was derived principally from his labors.

THE GOLD-BEATER, AND THE JEWELLER.

GOLD.

1. THE metals most extensively employed in the arts are gold, silver, copper, lead, tin, and iron. These are sometimes found uncombined with any other substance, or combined only with each other; in either of these cases, they are said to be in a *native state*. But they are more frequently found united with some substances which, in a great measure, disguise their metallic qualities, or, in other words, in a state of *ore*. The mode of separating the metals from their ores, will be noticed in connexion with some of the trades in which they are prepared for, or practically applied in, the arts.

2. Gold is a metal of a yellow color, a characteristic by which it is distinguished from all other simple

metallic bodies. As a representative of property, it has been used from time immemorial; and, before coinage was invented, it passed for money in its native state. In this form, gold is still current in some parts of Africa; and even in the Southern states of our own country, in the vicinity of the gold mines, the same practice, in a measure, prevails.

3. Gold is rarely employed in a state of perfect purity, but is generally used in combination with some other metal, which renders it harder, and consequently more capable of enduring the friction to which it is exposed. The metal used for this purpose is called an *alloy*, and generally consists of silver or copper.

4. For convenience in commerce, this precious metal is supposed to be divided into twenty-four equal parts, called *carats*. If perfectly pure, it is denominated gold 24 carats fine; if alloyed with one part of any other metal or mixture of metals, it is said to be 23 carats fine. The standard gold coin of the United States and Great Britain is 22 carats fine; or, in other words, it contains one-twelfth part of alloy. Gold, made standard by equal parts of copper and silver, approaches in color more nearly to pure gold than when alloyed in any other manner.

5. Gold is found in veins in mountains, most usually associated with ores of silver, sulphurets of iron, copper, lead, and other metals. It is often so minutely distributed, that its presence is detected only by pounding and washing the ores in which it exists. But the greatest part of the gold in the possession of mankind, has been found in the form of grains and small detached masses, amid the sands of rivers and in alluvial lands, where it had been deposited by means of water, which had detached it from its original position in the mountains.

6. To separate or extract gold from the foreign matters with which it may be combined, the whole is

first pounded fine, and then washed by putting it in a stream of water, which carries off the stony particles, while the gold, by its specific gravity, sinks to the bottom. To render the separation still more perfect, this sediment is mixed with ten times its weight of quicksilver, and put into a leather bag, in which it is submitted to a pressure that forces the fluid part through its pores; while the more solid part of the amalgam, which contains most of the gold, remains.

7. To separate the quicksilver from the gold, the mass is subjected to the process of *sublimation* in earthen retorts, which, as applied to metals, is similar in its effects to distillation, as applied to liquids. When gold is contained in the ores of other metals, they are roasted, in order to drive off the volatile parts, and to oxydize the other metals. The gold is then extracted by amalgamation, by liquefaction with lead, by the aid of nitric acid, or by other methods adapted to the nature of the ore.

8. Gold obtained in any of these methods is always more or less alloyed with some other metal, especially with silver or copper; but a separation is produced, so far as it is required for the purposes of commerce, by two processes, one of which is called *cupellation*, and the other *parting*. The former of these operations consists in melting the gold with a quantity of lead, which readily oxydizes and vitrifies, and which causes the same changes to take place in the metal to be detached from the mass of gold. The operation is called cupellation, because it is usually performed on a *cupel*, a vessel formed of bone-ashes, or sometimes of wood-ashes.

9. Cupellation is effectual in removing copper, but not so with regard to silver; the latter is separated by means of a process called *parting*. The metal is rolled out into thin sheets or strips, and cut into small pieces. These are put into diluted nitric acid, which,

by the aid of a moderate heat, dissolves the silver, leaving the gold in a porous state.

10. Another process, called *cementation,* is also sometimes used. It is performed by beating the alloyed metal into thin plates, and arranging them in alternate layers with a cement containing nitrate of potash, and sulphate of iron. The whole is then exposed to heat, until a great part of the baser metals has been removed by the action of the nitric acid liberated by the nitre. Cementation is often employed by goldsmiths, to refine the surface of articles in which the gold has been combined, in too small a proportion, with metals of less value.

11. The average amount of gold annually obtained in every part of the globe cannot fall far short of twenty-millions of dollars in value, of which South America supplies about one half, and Europe, about one twenty-fifth part. The amount yielded by the Southern states of our Union, cannot be accurately ascertained, but the whole sum coined at the United States' Mint in 1834, from gold obtained in this quarter, amounted to $898,000, and since 1824 to that time, to $3,679,000. In 1824, the sum was but $5000. Our Southern mines will probably continue to increase in productiveness.

THE GOLD-BEATER.

1. GOLD, not being subject to intrinsic change by atmospheric action, or by that of common chemical agents, is extensively used in gilding various substances, either with the view of preserving them from decay, or for the purpose of embellishment. To prepare the gold for application in this manner is the business of the gold-beater.

2. The metal is first melted with some borax in a crucible, and formed into an *ingot* by pouring it into an iron mould. The mass is next hammered a little

on an anvil, to increase the cohesion of its parts, and afterwards repeatedly passed between steel rollers, until it has become a riband as thin as paper.

3. Two ounces and a half of this riband are cut into 150 pieces of equal dimensions. These are hammered a little to make them smooth, and then interlaid with pieces of fine vellum four inches square. The whole, with twenty other pieces of vellum on each side, is inclosed in two cases of parchment. The packet is then beaten on a marble anvil with a hammer weighing sixteen pounds, until the gold has been spread to near the size of the vellum leaves, it, in the mean time, being often turned over.

4. The gold leaves are next divided into four equal squares, with a steel knife on a leather cushion; and the 600 leaves thus produced, are interlaid with a kind of leather or parchment made of the intestines of the ox, and beaten with a hammer weighing twelve pounds, until the leaves have been extended as before. They are again quartered and interlaid, and beaten with a hammer weighing six or eight pounds.

5. The gold having now been sufficiently extended, the packets are taken apart, and the leaves cut to a proper and uniform size, by means of a cane frame on a leather cushion. The leaves, as fast as they are trimmed, are placed in a book, the paper of which has been covered with red bole, to prevent the gold from sticking. Of the two ounces and a half of gold thus treated, only about one ounce remains in perfect leaves, which, altogether, amount to 2000 three inches and three-eighths square. The books contain twenty-five leaves, so that one ounce of gold makes eighty books.

6. Gold extended into leaves, is alloyed, in a greater or less degree, with silver or copper, or both, because, in a pure state, it would be too ductile. The newest skins will work the purest gold, and make the thinnest

leaf, because they are the smoothest. The alloy varies from three to twenty-four grains to the ounce, but in general it is six, or one part of alloy to eighty of gold.

7. A kind of leaf called *party gold*, is formed by the union of a thin leaf of gold and a thicker one of silver. The two are laid together, and afterwards heated and pressed, until they have cohered. They are then beaten and otherwise treated, as in the process just described. Silver, and likewise copper, are also beaten into leaves, although they will by no means bear so great a reduction as gold. Considerable quantities of copper leaf are brought from Holland, which in commerce is known by the denomination of "Dutch leaf," or "Dutch gold."

8. The ancient Romans were not ignorant of the process of gold-beating, although they did not carry it so far as we do. Pliny informs us that they sometimes made 750 leaves four fingers square, from an ounce of gold. At Præneste was a statue of Fortune, gilt with leaves of a certain thickness; hence those beaten to the same degree of thickness were called *Prænestines*. Those of another and less degree of thickness, were called *quæstoriales*, for a similar reason.

9. The Romans began to gild the interior of their houses immediately after the destruction of Carthage. The wainscots of the Capitol were first ornamented in this manner; and afterwards it became fashionable to gild the walls and ceilings of private dwellings, as well as articles of furniture.

10. *Gold wire.*—The ductility of gold is more conspicuous in wire than in leaves. The wire thus denominated, is in reality silver wire covered with gold. It is formed by covering a silver rod with thick leaves of gold, and then drawing it successively through conical holes of different sizes, made in plates of steel.

The wire may be reduced, in this manner, to a degree of extreme fineness, the gold being drawn out with the silver, and constituting for it a perfect coating.

11. Wire thus formed is often used in the manufacture of *gold thread.* Before it is applied in this way, it is flattened between rollers of polished steel, and then wound on yellow silk by machinery. The coating of gold on the silver wire employed in this way, does not exceed the millionth part of an inch in thickness.

THE JEWELLER.

1. THE jeweller makes rings, lockets, bracelets, brooches, ear-rings, necklaces, watch-chains, and trinkets of like nature. The materials of the best quality of these ornaments are gold, pearls, and precious stones, although those of an inferior kind are often used.

2. There are several stones to which is applied the epithet *precious,* of which the following are the principal: the diamond, the ruby, the sapphire, the topaz, the chrysolite, the beryl, the emerald, the hyacinth, the amethyst, the garnet, the tourmalin, and the opal. To these may be added rock crystal, the fine flints of pebbles, the cat's-eye, the oculis mundi or hydrophanes, the chalcedony, the moon-stone, the onyx, the carnelian, the sardonyx, agates, and the Labrador-stone. These stones, together with different kinds of pearl, are also called gems or jewels.

3. The precious stones are valuable, as articles of merchandise, in proportion to their scarcity, weight, transparency, lustre, and hardness. In most of these particulars, the diamond is superior to any other; but those of the same size are not always of equal value, for all are not of the same color or brilliancy. The very best are said to be *diamonds of the first water.* The diamond was called adamant by the an-

cients, although this term was not confined exclusively to this stone.

4. The weight and consequent value of the most precious stones are estimated in *carats*, one of which is equal to four grains troy weight, and the value of each carat is increased in proportion to the size of the stone. In England, the cost of a cut diamond of the first water is thus estimated:

$$\begin{array}{rl} 1\ \text{carat is} & =\ l.8 \\ 2\ \text{ do. is } 2\times2\times8 & =\ 32 \\ 3\ \text{ do. is } 3\times3\times8 & =\ 72 \\ 4\ \text{ do. is } 4\times4\times8 & =\ 128 \end{array}$$

By the foregoing examples, it will be seen that the weight is multiplied by itself, and the product by the price per carat, which may be some other sum, according to the general characteristics of the stone.

5. This rule, however, is not extended to diamonds of more than 20 carats in weight; nor is this or any other rule of estimate strictly adhered to in every case; nevertheless, it probably comes pretty near to general usage. In the same country, a perfect ruby of $3\frac{1}{2}$ carats is worth more than a diamond of equal weight. A ruby weighing one carat may be worth 10 guineas; two carats, 40 guineas; three carats, 150 guineas; six carats 1000 guineas. A ruby of a deep red color, exceeding 20 carats, is called a carbuncle; and of these, 108, weighing from 100 to 200 carats each, are said to have been in the throne of the Great Mogul.

6. Some of the European sovereigns have, in their possession, diamonds of great value, several of which were originally brought to England from India. The *Pitt* or *Regent diamond* was purchased in India by Robert Pitt, grandfather of the Right Honorable William Pitt, for £12,500 sterling. It was brought to England in a rough state, and £5000 were there expended in cutting it; but the cuttings themselves

were worth £7000 or £8000. It was sold to the Duke of Orleans, for the King of France, at the enormous price of £136,000. Its weight is 136 carats; and, before it was cut, it was as large as a common pullet's egg.

7. A celebrated diamond, in the possession of the emperor of Russia, is denominated the *Effingham* or *Russian diamond*. It was brought to England by the Earl of Effingham, while governor-general of India, and sold to the Empress Catharine for £90,000. It *is* inferior in shape to the last mentioned, but superior to it in magnitude, it weighing 198 carats. The Queen of England has a diamond which cost £22,000.

8. The largest diamond hitherto known was found in the island of Borneo, and it is now in the possession of the Rajah of Mattan. Many years ago, the governor of Batavia offered, in exchange for it, $150,000, and two large brigs of war with their equipments and outfit; but the rajah refused to part with the jewel, to which the Malays supposed miraculous power belonged, and which they believed to be connected with the fate of his family. The weight of this diamond is 367 carats.

9. Other jewels, belonging to different sovereigns, as well as to private persons, might be mentioned; but a sufficient number has been noticed to enable the reader to form some idea of the extravagant expenditures often made for articles of imaginary value. We will merely add that the royal family of Portugal is in possession of a stone which was formerly supposed to be a diamond, but which has lately proved to be some kind of crystal of little value. The weight of this stone is 1680 carats; and, until its real character was discovered, it was valued at 224 millions sterling

10. The value of precious stones was much increased in ancient times, by the absurd notion com-

monly entertained, that they possessed miraculous powers in preventing or curing diseases, as well as in keeping off witches and evil spirits. These notions still prevail more or less in heathen nations; and many, even in countries called Christian, wear them, or something else, as amulets for the same or similar purposes.

11. *The Gem-sculptor.*—Figures and letters are often cut in precious stones by the gem-engraver, or gem-sculptor, whose art, according to the opinion of some writers, originated with the Babylonians; but, according to others, it had its commencement in India or Egypt. In the latter country, it was first employed in the production of hieroglyphical figures on basalt and granite rocks. This art, which is denominated lithoglyptics, or the glyptic art, was held in great estimation by the Greeks in ancient times. It arose to eminence with the other fine arts; and, like them, it had its zenith of perfection, was buried with them in the ruins of the Roman empire, and with them revived towards the end of the fifteenth century.

12. The productions of gem-sculpture are chiefly of two kinds. The first of these are *cameos*, which are little bas-reliefs, or figures raised above the surface. They are commonly made of stones, the strata of which are of different colors, so that the figure is different in color from the ground on which it has been raised. The other productions of this art are denominated *intaglios*. The work of these is the reverse of that first mentioned, since the figure is cut below the surface of the stone, so that they serve as seals to produce impressions in relief upon soft substances.

13. This artist performs his work by means of a lathe, with the aid of diamond dust. The instruments are made of soft iron, and are fixed in leaden chucks, which can be readily fastened to one end of

the mandi. The diamond dust is made into thin paste with on e oil, and is applied to the point of the instrument. The small invisible particles insinuate themselves into the iron, where they remain permanently fixed. In producing figures and letters with a tool thus charged with the hardest substance in nature, the precious stone is brought in contact with it while in rapid motion.

14. The engraved gems of antiquity have been greatly esteemed, as works of art, by the curious, and various methods have, therefore, been devised to imitate them. This has been done in glass in such perfection, both as to form and color, that good judges can scarcely distinguish the imitations from the originals. The impression of the gem is first taken in some kind of fine earth; and, upon the mould thus formed, the proposed material is pressed, while in a plastic state.

15. The precious stones generally have likewise been imitated with great success. The basis of the different compositions is a *paste* made of the finest flint glass, the materials of which have been selected and combined with great care. The desired color is produced with metallic oxydes. A great number of complex receipts are in use among manufacturers of these articles.

16. *The Lapidary.*—The precious stones and imitations of them in glass are brought to the desired form by the lapidary. The instrument with which this artist chiefly operates is a wheel which is made to revolve horizontally before him. It is put in motion by means of an endless rope extending from another wheel, which is moved with the left hand of the operator, while, with his right, he holds, in a proper position, the substance to be reduced.

17. The precious stones, being of small size, cannot be held with steadiness on the wheel with the

hand, nor with any holding instrument; they are, therefore, first fastened, by means of sealing-wax, to the end of small sticks. By this simple means, and a small upright post, against which the hand or the other end of the stick is rested, the workman can hold a stone in any position he may desire.

18. The lapidary's wheel is made of different kinds of metals. The diamond is cut on a wheel of soft steel, by the aid of its own dust mixed with olive oil. The Oriental ruby, sapphire, and topaz, are cut on a copper wheel in the same manner, and polished with tripoli and water. Stones of a less degree of hardness are cut and polished on a leaden or tin wheel with emory and rotten stone.

19. The ancients were not acquainted with any method of cutting the diamond, although they applied its powder to polishing, cutting, and engraving other stones. Gems of this kind, either rough, or polished by nature, were set as ornaments, and were valued according to the beauty and perfection of their crystallization and transparency. The value of any precious stone, or jewel, depends much upon the skill of the lapidary.

20. *The Pearl-fisherman.* — Pearls are obtained from a testaceous fish of the oyster kind, found in the waters of the East and West Indies, as well as in other seas of different latitudes. These oysters grow in some parts of the globe, in clusters, on rocks in the depths of the sea. Such places are called *pearl-banks,* of which the most celebrated are near the islands of Ceylon and Japan, and in the Persian Gulf. The finest and most costly pearls are the Oriental.

21. Pearls are considered by some to be morbid concretions, or calculi, produced by the endeavor of the animal to fill up the holes which may have been made from without by small worms. Others suppose them to be mere concretions of the animal juice about

II.—R

some extraneous matter which may have been intruded by some means into the shell.

22. To collect the shells containing these singular productions, is the business of *divers*, who have been brought up to this dangerous occupation. They must generally descend from eight to twelve fathoms, and must remain beneath the surface of the water for several minutes, during which time they are exposed to the attacks of the voracious shark. In addition to the danger from this cause, the employment is very destructive of health.

23. In preparing a diver for his descent, a rope is tied round the body, and a stone, weighing twenty or thirty pounds, is fastened to the foot to sink him. His ears and nostrils are filled with cotton, and a sponge dipped in oil is fastened to his arm, to which he may now and then apply his mouth, in order to breathe without inhaling water. In addition to these equipments, he is furnished with a knife, with which the shells may be detached from the rocks, and with a net or basket, in which they may be deposited.

24. Thus equipped, he descends to the bottom, and having filled his depository, or having stayed below as long as he may be able, he unlooses the stone, gives the signal to his companions above, who quickly draw him into the boat. At some pearl-fisheries, the diving-bell is employed, which in some degree obviates some of the dangers before stated.

25. The shells thus obtained are laid by, until the body of the animal has putrified, when they commonly open of themselves. Those which contain any pearls, generally have from eight to twelve. The pearls having been dried, are assorted according to their various magnitudes; and, to effect this separation, they are passed through nine sieves of different degrees of fineness. The largest pearls are about

the size of a small walnut; but such are very rare. The smallest are called *seed pearls*.

26. Pearls are of various colors, such as white, yellow, lead-color, blackish, and totally black. The "white water" is preferred in Europe, and the "yellow water," in Arabia and India. In regard to their form, they vary considerably, being round, pear-formed, onion-formed, and irregular. The inner part of the pearl muscle is called *nacre* or *mother of pearl*, and this is manufactured into beads, snuff-boxes, spoons, and a variety of other articles.

27. Pearls were objects of luxury among the ancients. A pearl valued by Pliny at a certain sum, which, reduced to our currency, amounts to $375,000, was dissolved by Cleopatra, and drunk to the health of Antony, at a banquet. These beautiful productions are not estimated so highly at present. The largest will sometimes command four or five hundred dollars, although very few, which are worth over forty or fifty dollars, are ever brought to this country.

28. The gem-engraver and the jeweller were both employed by Moses, in preparing the ornaments in the ephod and breast-plate of the high-priest. In the former were set onyx stones, and in the latter, twelve different stones. On the gems of both ornaments, were engraved the names of the twelve tribes of Israel.

39. We, however, have evidence of the practice of the arts, connected with the production of jewelry, long before the days of the Jewish lawgiver. We learn from the twenty-fourth chapter of Genesis, that the servant of Abraham presented a golden ear-ring, and bracelets for the hands, to Rebecca, who afterwards became the wife of Isaac. Perhaps these were brought from Egypt by the patriarch, about seventy years before.

30. Men have ever been fond of personal ornaments, and there have been but few nations since the flood,

which have not encouraged the jeweller in some way or other. In modern times, the art has been greatly improved. The French, for lightness and elegance of design, have surpassed other nations; but the English, for excellence of workmanship, have been considered, for ages, unrivalled.

31. In the United States, the manufacture of jewelry is very extensive, there being large establishments for this purpose in Philadelphia, and in Newark, N. J., as well as in several other places. So extensive have been the operations in this branch of business, and to such advantage have they been carried on, that importations from other countries have ceased, and this, too, without the influence of custom-house duties.

32. The capital necessary in carrying on the business of the jeweller, is considerable, inasmuch as the materials are very expensive. The operations likewise require the exercise of much ingenuity. These, however, we shall not attempt to describe, since our article on this subject has already been extended beyond its proper limits, and since, also, they could be hardly understood without actual inspection.

THE SILVERSMITH, AND THE WATCH-MAKER.

SILVER.

1. SILVER is a metal of a fine white color, and, in brilliancy, inferior to none of the metals except steel. In malleability, it is next to gold, it being capable of reduction into leaves not more than the $\frac{1}{100000}$ of an inch in thickness, and of being drawn into wire much finer than a human hair.

2. The relative value of silver and gold has varied considerably in different ages. In the prosperous period of ancient civilization, one pound of gold was worth twelve of silver. In Great Britain, the relative value of the two metals is one to fifteen and one-fifth and, on the continent of Europe, it is about one to fifteen. In the United States, the relative value of these two metals has been recently established by Congress

at one to sixteen. In China and Japan, it is said to be one to nine or ten.

3. There are two methods of separating silver from its various ores, and these are called *smelting* and *amalgamation*. In the former method, the ore and a due proportion of lead are heated together; and the latter, from its great affinity for silver, unites with it, and separates it from other substances. The two metals are afterwards separated from each other, by melting them on a cupel, and then exposing them to a current of atmospheric air, by which the lead is converted into an oxyde, while the silver remains untouched. This process is called *cupellation*.

4. In the other method, the first thing done is to roast the ore, to expel the sulphur and other volatile parts. It is then reduced to an impalpable powder by machinery; and having been sifted, it is agitated sixteen or eighteen hours in barrels, with a quantity of quicksilver, water, and iron, combined in certain proportions. This agitation causes the several substances composing the *charge*, to unite according to their respective affinities.

5. The silver and mercury combine, forming an amalgam, which, having been put into a leather sack, a part of the latter is separated from the rest by filtration, still leaving six parts of this metal to one of the silver. The amalgam is next submitted to the action of heat in a distilling furnace, by which the mercury is sublimated.

6. The value of the silver annually taken from the mines in all parts of the world, is supposed to be about $20,000,000, of which Mexico and South America yield the greater part. The several silver mines of Europe and Asia produce about two millions and a half.

THE SILVERSMITH.

1. THE artisan who forms certain articles of gold and silver, is called indifferently a goldsmith or a silversmith. The former denomination is most commonly employed in England, and the latter, in the United States.

2. The most common subjects of manufacture by the silversmith are cups, goblets, chalices, tankards, spoons, knives, forks, waiters, bread-trays, tea-pots, coffe-pots, cream-pots, sugar-bowls, sugar-tongs, and pencil-cases. Many of these articles he sometimes makes of gold; this is especially the case in Europe, and some parts of Asia. In the United States, the people are commonly satisfied with the less expensive metal.

3. A great proportion of the silver used by this mechanic, has been previously coined into dollars. In working these into different utensils or vessels, he first melts them in a crucible, and casts the silver into solid masses by pouring it into iron moulds; and having forged it on an anvil, he reduces it still further, and to a uniform thickness, by passing it several times between steel rollers. In giving additional explanations of the operations of the silversmith, we will describe the manner in which a plain tea-pot is manufactured.

4. In forming the body, or containing part, the plate, forged and rolled as just described, is cut into a circular form, and placed on a block of soft wood with a concave face, where it is beaten with a convex hammer, until it has been brought to a form much like that of a saucer. It is then placed upon an anvil, and beaten a while with a long-necked hammer with a round flattish face.

5. It is next *raised* to the proposed form by forging it on a long slender anvil, called a *stake*, with a nar-

row-faced hammer, which spreads the metal perpendicularly from the bottom, or laterally, according to the position in which it may be held when brought in contact with the metal.

6. After the piece has been thus brought to the proposed form, it is *planished* all over by beating it with a small hammer on the outside, while it rests on a small steel head on the inside. During the performance of these operations, the silver is occasionally *annealed* by heating it in the fire; but it is worked while in a cold state, except in the first forging, when it is wrought while a little below red heat.

7. The several pieces which compose a tea-pot of ordinary construction, amount to about fifteen, nearly all of which are rolled and forged in the manner just described. The knob on the lid, the handle, and the spout, are sometimes cast, and at other times, the two pieces of which they are formed are cut from a plate, and brought to a proper figure by impressing them with steel dies.

8. The figures seen on the cheaper kinds of silver tea-pots, as well as on other vessels and utensils, are commonly made by passing the plates or strips between engraved steel rollers, or by stamping them with steel dies. The dies are commonly brought in sudden and violent contact with the metal by means of an iron *drop*, which is let fall from a height upon it.

9. After the several parts have been brought to the proper shape, and to the requisite finish, they are firmly united together by means of a solder composed of about three parts of silver and one of brass and copper. Before the spout and handle are soldered on, the other parts, which have been thus united into one piece, are brought to a certain degree of polish.

10. This is effected chiefly in a lathe, by holding against the piece, while in rapid motion, first a file, then a scraper, and afterwards pumice stone and

Scotch stone. It is then held against a rapidly revolving brush, charged with fine brickdust and sweet oil. The handle and spout are next soldered on. After this, the vessel is annealed, and put in *pickle*, or, in other words, into a weak solution of oil of vitriol. It is then scoured with sand and water, and the whole operation is completed by burnishing the smooth parts with a steel instrument.

11. In the more expensive kinds of wares, the raised figures and the frosty appearance are produced by a process called *chasing*. In executing this kind of work, a drawing is first made on the silver with a lead pencil. The several parts are then raised from the other side, corresponding as nearly as possible to it. The vessel or piece is then filled with, or placed upon, melted cement, composed of pitch and brick-dust; and, after the cement has become hard by cooling, the chaser reduces the raised parts to the form indicated by the drawing, by means of small steel punches. The roughness of surface, and frosty appearance, are produced by punches indented on the end.

12. The operations of the silversmith are exceedingly various, many of which could be hardly understood from mere description. We would, therefore, recommend to the curious, actual inspection, assuring them that the ingenuity displayed in executing the work in the different branches of the business, is well worthy of their attention. We will merely add, that spoons, knives, and forks, are not cast, as is frequently supposed, but forged from strips of silver cut from rolled sheets.

13. The earliest historical notice of gold and silver is found in the thirteenth chapter of Genesis, where it is stated that Abraham returned to Canaan from Egypt, "rich in cattle, in silver, and in gold." This event took place about 1920 years before Christ, it being but little more than 400 years after the deluge. From

the authority of the same-book, we also learn, that during the life of this patriarch, those metals were employed as a medium of commercial intercourse, and as the materials for personal ornaments, vessels, and utensils.

14. From the preceding facts, we have reason to believe that gold and silver were known to the antedi-luvians; for, had not this been the case, they could hardly have been held in such estimation so early as the time of Abraham. In short, they were probably wrought even in the days of the original progenitor of the human race, as was evidently the case with iron and copper.

THE CLOCK AND WATCH MAKER.

1. THE great divisions of time, noted by uncivilized men, are those which are indicated by the changes of the moon, and the annual and diurnal revolutions of the earth; but the ingenuity of man was very early exercised in devising methods of measuring more minute periods of duration.

2. The earliest contrivance for effecting this object was the sun-dial. This instrument was known to the ancient Egyptians, Chaldeans, Chinese, and Bramins. It was likewise known to the Hebrews, at least as early as 740 years before Christ, in the days of Ahaz the king. The Greeks and the Romans borrowed it from their Eastern neighbors. The first sun-dial at Rome was set up by Papirius Cursor, about 300 years before Christ. Before this period, the Romans determined the time of day by the rude method of observing the length of shadows.

3. The sun-dial, as it is now constructed, consists of a plate, divided into twelve equal parts, like the face of a clock, on which the falling of a shadow indicates the time of day. The shadow is projected by the sun, through the intervention of a rod or the edge

of a *plate stile* erected on the plane of the dial. But, since the dial was useful only in the clear day, another instrument was invented, which could be used at all times, in every variety of situation; and to this was given the name of *clepsydra.*

4. This instrument is supposed to have been invented in Egypt; but, at what period, or by whom, it is not stated. Its construction was varied, in different ages and countries, according with the particular modes of reckoning time ; but the constant dropping or running of water from one vessel into another, through a small aperture, is the basis in all the forms which it has assumed. The time was indicated by the regularly increasing height of the water in the receiving vessel.

5. The clepsydra was introduced into Greece by Plato, near 400 years before Christ, and, about 200 years after this, into Rome, by Scipio Africanus. It is said that Pompey brought a valuable one from the East, and that Julius Cæsar met with one in England, by which he discovered that the summer nights were shorter there than in Italy.

6. The use which Pompey made of his instrument, was to limit the length of speeches in the senate. Hence he is said, by a historian of those times, to have been the first Roman who put bridles upon eloquence. A similar use was made of the clepsydra in the courts of justice, first in Greece, and afterwards in Rome.

7. A kind of water-clock, or clepsydra, adapted to the modern divisions of time, was invented near the middle of the seventeenth century; and these were extensively used, in various parts of Europe, for a considerable time ; but they are now entirely superseded by our common clocks and watches, which are far more perfect in their operation, and, in all re-

spects, better adapted to the purposes to which they are applied.

8. The invention of the clock is concealed in the greatest obscurity. Some writers attribute it to the monks, as this instrument was used in the twelfth century in the monasteries, to regulate the inmates in their attendance on prayers both by night and by day. · Others suppose that a knowledge of this valuable instrument was derived from the Saracens, through the intercourse arising from the crusades. Be this as it may, clocks· were but little known in Europe, until the beginning of the fourteenth century.

9. Richard, abbot of St. Alban's, England, made a clock in 1326, such as had never been heard of until then. It not only indicated the course of the sun and moon, but also the ebbing and flowing of the tide. Large clocks on steeples began to be used in this century. The first of this kind is supposed to have been made and put up in Padua by Jacobus Dondi.

10. A steeple clock was set up in Boulogne, in 1356 ; and, in 1364, Henry de Wyck, a German artist, placed one in the palace of Charles V., king of France. In 1368, three Dutchmen introduced clockwork into England, under the patronage of Edward III. Clocks began to be common both in England and on the Continent, about the end of the fifteenth century.

11. The clock of Henry de Wyck is the most ancient instrument of this kind of which we have a description. The wheels were made of wrought iron, and the teeth were cut by hand. In other respects, also, it was a rude piece of mechanism, and not at all capable of keeping time with accuracy. But, rude as it was, it is not likely that it was the invention of a single individual; but that, after the first rude machine was put in motion, it received several improvements from various persons. This has, at least, been

the case with all the improvements made on the clock of Henry de Wyck, to the present day.

12. The application of the pendulum to clock-work appears to have been first made by Vincenzo Galileo, in 1649; but the improvement was rendered completely successful, in 1656, by Christian Huygens, a Dutch philosopher. The laws of the oscillation of the pendulum were first investigated by Galileo, the great Italian philosopher, and father of the Galileo just mentioned. His attention was attracted to this subject by the swinging of a lamp suspended from the ceiling of the Cathedral, at Pisa, his native city.

13. The clocks first made were of a large size, and were placed only in public edifices. The works were, at length, reduced in their dimensions, and these useful machines were gradually introduced into private dwellings. They were finally made of a portable size, and were carried about the person. These portable clocks had, for their maintaining power, a mainspring of steel, instead of a weight, which was used in the larger time-keepers.

14. The original pocket-watches differed but little, if at all, in the general plan of their construction, from the portable clocks just mentioned. The transition from one kind of instrument to the other was, therefore, obvious and easy; but the time of the change cannot be certainly determined. It is commonly admitted, however, that Peter Hele constructed the first watch, in 1510.

15. Watches appear to have been extensively manufactured at Nuremburg, in Germany, soon after their invention, as one of the names by which they were designated, was *Nuremburg eggs*. These instruments, as well as clocks, were in common use in France, in 1544, when the company of clock and watch makers of Paris was first incorporated.

16. In 1658, the spring balance was invented by

Doctor Nathaniel Hooke, an English philosopher. At least the invention is attributed to him by his countrymen. On the Continent it is claimed for Christian Huygens. Before this improvement was made, the performance of watches was so defective, that the best of them could not be relied upon for accurate time an hour together. Their owners were obliged to set them often to the proper time, and wind them up twice a day.

17. After the great improvements had been effected in the clock and watch by Huygens and Hooke, several others of minor importance were successively made by different persons; but our limits do not allow us to give them a particular notice; we will only state that the repeating apparatus of both clocks and watches was invented, about the year 1676, by one Barlow, an Englishman; that the compensation or gridiron pendulum was invented by George Graham, of London, in 1715; and that jewels were applied to watches, to prevent friction, by one Facio, a German.

18. Clocks and watches are constructed on the same general principles. The mechanism of both is composed of wheel-work, with contrivances to put it in motion, and to regulate its movements. The moving or maintaining power in large clocks is a weight suspended by a cord to a cylinder. In watches, and sometimes in small clocks, this office is performed by a steel spring. In the clock, the regulation of the machinery is effected by the pendulum, and in the watch, by the balance-wheel, or spring balance. In either case, the maintaining power is prevented from expending itself, except in measured portions.

19. The time is indicated by hands, or pointers, which move on the dial plate. The minute hand is attached to the axle of the wheel which makes its revolution in sixty minutes, and the hour hand, to the one which makes the revolution in twelve hours. Great-

er and smaller divisions of time are kept and indicated on the same principle. The part of a clock which keeps the time, is called the going part; and that which strikes the hour, the striking part.

20. The division of labor is particularly conspicuous in the manufacture of watches, as the production of almost every part is the labor of a distinct artisan. The workman who polishes the several parts, and puts them together, is called, among this class of tradesmen, the *finisher* or *watch-maker*. Those, therefore, who deal largely in watches in England, purchase the different parts from the several manufacturers, and cause them to be put together by the finisher.

21. Watches are extensively manufactured in various parts of Europe, but particularly in French Switzerland, France, and England. The London watchmakers have been celebrated for good workmanship, for more than a century and a half. This manufacture has not yet been commenced in the United States, although the machinery, or *inside work*, is very often imported in tin boxes, and afterwards supplied with dial plates and cases. This is especially the case with the more valuable kinds of watches.

22. Brass clocks are maufactured in most of our cities, and in many of our villages, and wooden clocks, in great numbers, in the state of Connecticut. These last are carried by pedlers into the remotest parts of the country, so that almost every farmer in our land can divide the day by the oscillations of the pendulum.

THE COPPERSMITH, THE BUTTON-MAKER, AND THE PIN-MAKER.

COPPER.

1. COPPER is a ductile and malleable metal, of a pale yellowish red color. It is sometimes found in a native state, but not in great quantities. The copper of commerce is principally extracted from the ores called sulphurets. Copper mines are wrought in many countries; but those of Sweden are said to furnish the purest copper of commerce, although those of the island of Anglesea are said to be the richest.

2. In working sulphureted ore, it is first broken into pieces, and roasted with a moderate heat in a kiln, to free it from sulphur. When the ore is also largely combined with arsenic, a greater degree of heat is necessary. In such a case, it is spread upon a large

floor of a reverberatory furnace, and exposed to a greater heat. By this treatment, the sulphur and arsenic are soon driven off.

3. The ore is then transferred to the fusing furnace, and smelted in contact with fuel. The specific gravity of the copper, causes it to sink beneath the *scoria* into a receptacle at the bottom of the furnace. To render the metal sufficiently pure, it requires repeated fusions, and, even after these, it usually contains a little lead, and a small portion of antimony.

4. *Alloys of copper.*—Copper is combined by fusion with a great number of metals, and, in such combinations, it is of great importance in the arts. When added in small quantities to gold and silver, it increases their hardness, without materially injuring their color, or diminishing their malleability. An alloy, called white copper, imported from China, and denominated, in that country, *pakfong*, is composed of copper, zinc, nickel, and iron. It is very tough and malleable, and is easily cast, hammered, and polished. When well manufactured, it is very white, and as little liable to oxydation as silver.

5. Copper, with about one-fourth of its weight of lead, forms *pot-metal*. *Brass* is an alloy of copper and zinc. The proportion of the latter metal varies from one-eighth to one-fourth. Mixtures, chiefly of these two metals, are also employed to form a variety of gold-colored alloys, among which are *prince's metal*, *pinchbeck*, *tombac*, and *bath-metal*.

6. A series of alloys is formed by a combination of tin and copper. They are all more or less brittle, rigid, and sonorous, according to the relative proportions of the two metals; these qualities increasing with the amount of tin. The principal of these alloys are, *bronze*, employed in the casting of statues; *gun-metal*, of which pieces of artillery are made; *bell-metal*, of which bells are made; and *speculum-*

metal, which is used for the mirrors of reflecting telescopes.

7. The alloys of copper were very prevalent among the nations of antiquity, and were used, in many cases where iron would have answered a much better purpose. The instruments of husbandry and of war, as well as those for domestic uses generally, were usually made of bronze, a composition which furnishes the best substitute for iron and steel. The Corinthian brass, so celebrated in antiquity, was a mixture of copper, gold, and silver.

8. The earliest information of the use of this metal by mankind, is found in the fourth chapter of Genesis, in which it is stated, that "Tubal-Cain was the instructer of every artificer in brass and iron." This individual was the seventh generation from Adam, and was born about the year of the world 500.

THE COPPERSMITH.

1. COPPER, being easily wrought, is applied to many useful purposes. It is formed into sheets by heating it in a furnace, and compressing it between steel rollers. The operation of rolling it constitutes a distinct business, and is performed in mills erected for the express purpose.

2. The rolled sheets are purchased according to weight by the coppersmith, who employs them in sheathing the bottoms of ships, in covering the roofs of houses, and in constructing steam-boilers and stills. He also fabricates them into a variety of household utensils, although the use of this metal in preparing and preserving food, is attended with some danger, on account of the poisonous quality of the verdigris which is produced on the surface.

3. An attempt has been made to obviate this difficulty, by lining the vessels with a thin coating of tin. This answers the purpose fully, so long as the cover-

ing of tin remains entire. But, in cases of exposure to heat, it is liable to be melted off, unless it is kept covered with water.

4. This metal can be reduced by forging to any shape; but, during the process, it will bear no heat greater than a red heat; and, as it does not admit of welding, like iron, different pieces are united with bolts, or rivets, of the same metal, as in the case of the larger kinds of vessels, or by means of solder made of brass and zinc, or zinc and lead, as in the case of those of smaller dimensions.

5. Brass is applied to a greater variety of purposes in the arts than copper. This preference has arisen from its superior beauty, from the greater facility with which it can be formed into any required shape, and from its being less influenced by exposure to the ordinary chemical agents.

6. Some of the articles manufactured of brass, are forged to the required form, and others are made of rolled sheets; but, in most cases, they pass through the hands of the brass-founder, who liquifies the metal, and pours it into moulds of sand. For the sake of lightness, and economy of material, many articles are made hollow; in such cases, they are cast in halves or pieces, and these are afterwards soldered together.

7. Pieces which have been cast are generally reduced in size, and brought more exactly to the proposed form, either in a lathe, with tools adapted to turning, or in the vice, with files and other suitable instruments. The operators in brass form a class of mechanics distinct from those who work in copper.

THE BUTTON-MAKER.

1. TRIFLING as the manufacture of buttons may appear, there are few which include a greater variety of operations. The number of substances of which they are made is very great, among which are gold,

silver, various alloys of copper, steel, tin, glass, mother of-pearl, bone, horn, and tortoise-shell, besides those which consist of moulds of wood or bone, covered with silk, mohair, or similar materials.

2. In making gilt buttons, the *blanks*, or bodies, are cut from rolled plates of brass, with a circular punch driven by means of a fly wheel. The blanks thus produced, are planished with a plain die, if they are intended for plain buttons; or with one having on it an engraved figure, if they are to be of the ornamental kind. In either case, the die is usually driven with a fly press.

3. The shanks are next placed on one side of the proposed button, and held there temporarily with a wire clasp. A small quantity of solder and rosin having been applied to each shank, the buttons are exposed to heat on an iron plate, until the solder shall have melted. The shanks having been thus firmly soldered on, the buttons are turned off smoothly on their edges in a lathe.

4. The buttons are next freed from oxyde, by immersing them in diluted nitric acid, and by friction in a lathe. They are then put into a vessel containing a quantity of nitric acid supersaturated with mercury. The superior attraction of the copper for the acid, causes a portion of it to be absorbed; and the mercury held in solution by it, is deposited on the buttons, which are next put into a vessel containing an amalgam of mercury and gold.

5. The amalgam is formed by melting the two metals together, and afterwards pouring them into cold water. The composition having been put into a bag of chamois leather, and a part of the mercury pressed through the pores, the remaining portion is left in a condition approaching the consistency of butter, and in a fit state for use. Before the buttons are

put into the amalgam, a small quantity of nitric acid is added.

6. The buttons having been covered with the amalgam, as before stated, the mercury is discharged, that the gold may adhere directly to the brass. This object is effected by heating the buttons in an iron pan, until the amalgam begins to melt, when they are thrown into a large felt cap, and stirred with a brush. This operation is repeated several times, until all the mercury has been volatilized. The whole process is finished by again burnishing them, and putting them on cards for sale.

7. White metal buttons are made of brass alloyed with different proportions of tin. They are cast, ten or twelve dozens at a time, in moulds formed in sand, by means of a pattern. The shanks are placed in the centre of the moulds, so that, when the metal is poured in, they become a part of the buttons. The buttons are next polished in a lathe, with grindstone dust and oil, rotten stone and crocus martis. They are then boiled with a quantity of grained tin, in a solution of crude red tartar or argol, and lastly, finished with finely-pulverized crocus, applied with buff leather.

8. Glass buttons are made of various colors, in imitation of the opal and other precious stones. While manufacturing them, the glass is kept in a state of fusion, and a portion of it for each button is nipped off out of the crucible with a metallic mould, somewhat similar to that used for running bullets, the workman having previously inserted into it the shank.

THE PIN-MAKER.

1. THERE is scarcely any commodity cheaper than pins, and none which passes through the hands of a greater number of workmen in the manufacture, twenty-five persons being successively employed upon the

material, before it appears in these useful articles, ready for sale.

2. The wire having been reduced to the required size, is cut into pieces long enough to make six pins. These pieces are brought to a point at each end by holding them, a handful at a time, on a grindstone. This part of the operation is performed with great rapidity, as a boy twelve years of age can sharpen 16,000 in an hour. When the wires have been thus pointed, the length of a pin is taken off at each end, by another hand. The grinding and cutting off are repeated, until the whole length has been used up.

3. The next operation is that of forming the heads, or, as the pin-makers term it, *head-spinning*. This is done with a *spinning-wheel*, by which one piece of wire is wound upon another, the former, by this means, being formed into a spiral coil similar to that of the springs formerly used in elastic suspenders. The coiled wire is cut into suitable portions with the shears, every two turns of it being designed for one head. These heads are fastened to the *lengths* by means of a hammer, which is put in motion with the foot, while the hands are employed in taking up, adjusting, and placing the parts upon the anvil.

4. The pins are now finished, as to their form; but still they are merely brass. To give them the requisite whiteness, they are thrown into a copper vessel, containing a solution of tin and the lees of wine. After a while, the tin leaves the liquid, and fastens on the pins, which, when taken out, assume a white appearance. They are next polished by agitating them with a quantity of bran in a vessel moved in a rotary manner. The bran is separated from them, as chaff is separated from wheat.

5. Pins are also made of iron wire, and colored black by a varnish composed of linseed oil and lampblack. This kind is designed for persons in mourn-

ing. Pins are likewise made with a head at each end, to be used by females in adjusting the hair for the night, without the danger of pricking. Several machines have been invented for this manufacture, one of which makes a solid head from the body of the pin itself; but the method just described still continues to be the prevailing one.

6. Pins are made of various sizes. The smallest are called minikins, the next, short whites. The larger kinds are numbered from three to twenty, each size increasing one half from three to five, one from five to fourteen, and two from fourteen to twenty. They are put up in papers, according to their numbers, as we usually see them, or in papers containing all sizes. In the latter case, they are sold by weight.

7. It is difficult, or even impossible, to trace the origin of this useful little article. It is probable, however, that it was invented in France, in the fifteenth century. One of the prohibitions of a statute, relating to the pin-makers of Paris of the sixteenth century, forbid any manufacturer to open more than one shop for the sale of his wares, except on new-year's day, and on the day previous.

8. Hence we may infer, that it was customary to give pins as new-year's presents, or that it was the usual practice to make the chief purchases at this time. At length it became a practice, in many parts of Europe, for the husband to allow to his wife a sum of money for this purpose. We see here the origin of the phrase, *pin-money*, which is now applied to designate the sum allowed to the wife for her personal expenses generally.

9. Prior to the year 1443, the art of making pins from brass wire was not known in England. Until that period, they were made of bone, ivory, or boxwood. Brass pins are first mentioned in the English statute book, in 1483, when those of foreign manufacture were prohibited

10. Although these useful implements are made in London, and in several other places in England, yet Gloucester is the principal seat of this manufacture in that kingdom. It was introduced into that place, in 1626, by John Silsby, and it now contains nine distinct manufactories, in which are employed about 1500 persons, chiefly women and children. Pins are also manufactured extensively in the villages near Paris, and in several other places in France, as well as in Germany.

11. The business of making pins has been lately commenced in the city of New-York, and it is said that the experiment has been so successful, both in the perfection of the workmanship, and in the rapidity of the production, that pins of American manufacture bid fair to compete, at least, with those of foreign countries.

THE TINPLATE WORKER, &c.

TIN.

1. TIN is a whitish metal, less elastic, and less sonorous than any other metal, except lead. It is found in the mountains which separate Gallicia from Portugal, and in the mountains between Saxony and Bohemia. It also occurs in the peninsula of Molucca, in India, Mexico, and Chili. But the mines of Cornwall and Devonshire, in England, are more productive than those of all other countries united.

2. There are two ores of tin, one of which is called *tin stone*, and the other *tin pyrites;* the former of these is the kind from which the metal is extracted. The ore is usually found in veins, which often penetrate the hardest rocks. When near the surface of the earth, or at their commencement, they are very small,

but they increase in size, as they penetrate the earth. The direction of these veins, or, as the miners call them, *lodes*, is usually east and west.

3. The miners follow the lode, wheresoever it may lead; and, when they extend to such a depth, that the waters become troublesome in the mine, as is frequently the case, they are pumped up with machinery worked by steam, or drawn off by means of a drain, called an *adit*. The latter method is generally adopted, when practicable.

4. The ore is raised to the surface through shafts, which have been sunk in a perpendicular direction upon the vein. At the top of the shaft, is placed a windlass, to draw up the *kibbuts*, or baskets, containing the ore. Near St. Austle, in Cornwall, is a mine which has not less than fifty shafts, half of which are now in use. Some of the veins have been worked a full mile, and some of the shafts are nearly seven hundred feet deep.

5. At St. Austle Moor, there is a mine of *stream tin*, about three miles in length. The tin, together with other substances, has been deposited in a valley, by means of small streams from the hills. The deposite is about twenty feet deep, and the several materials of which it is composed, have settled in strata, according to their specific gravity. The ore, being the heaviest, is, of course, found at the bottom.

6. The ore, from whatever source it may be obtained, is first pulverized in a stamping mill, and then washed, to free it from the stony matter with which it may be united. The ore, thus partially freed from foreign matter, is put into a reverberatory furnace, with fuel and limestone, and heated intensely. The contents of the furnace having been brought to a state of fusion, the lime unites with the earthy matters, and flows with them into a liquid glass, while the carbon of the coal unites with the tin. The metal sinks, by

its specific gravity, to the bottom of the furnace, and is let out, after having been exposed to the heat about ten hours.

7. The tin thus obtained, is very impure; it therefore requires a second fusion, to render it fit for use. After having been melted a second time, it is cast into blocks weighing about three hundred pounds. These blocks are taken to places designated by law, and there stamped, by inspectors appointed for the purpose by the Duke of Cornwall. In performing this operation, the inspector cuts off a corner, and stamps the block at that place, with the proper seal, and with the name of the smelter. These precautions give assurance, that the metal is pure, and that the duty has been paid.

8. The duty is four shillings sterling per hundred weight, which is paid to the Duke of Cornwall, who is also Prince of Wales. The revenue from this source amounts to about thirty thousand pounds a year. The owner of the soil also receives one sixth, or one eighth of the ore as his *dish*, as the miners call it. The miners and the smelters receive certain proportions of the metal for their services.

9. Tin was procured from Britain at a very early period. The Phœnicians are said by Strabo to have passed the Pillars of Hercules, now the Straits of Gibraltar, about 1200 years before Christ. But the time at which they discovered the tin islands, which they denominated *Cassorides*, cannot be ascertained from history, although it is evident from many circumstances, that the Scilly Islands, and the western ports of Britain, were the places from which these early navigators procured the tin with which they supplied the parts of the world to which they traded.

10. For a long time, the Phœnicians and the Carthaginians enjoyed the tin trade, to the exclusion of all other nations. After the destruction of Carthage by

the Romans, a colony of Phocean Greeks, established at Marseilles, carried on this trade; but it came into the hands of the Romans, after the conquest of Britain by Julius Cæsar.

11. The Cornish mines furnish incontestable proofs of having been worked many hundred years ago. In digging to the depth of forty or fifty fathoms, the miners frequently meet with large timbers imbedded in the ore. Tools for mining have also been found in the same, or similar situations. The mines, therefore, which had been exhausted of the ore, have, in the course of time, been replenished by a process of nature.

12. To what purposes the ancients applied all the tin which they procured at so much labor and cost, is not precisely known. It is probable, that the Tyrians consumed a portion of it, in dyeing their purple and scarlet. It formed then, as it now does, many important alloys with copper. The mirrors of antiquity were made of a composition of these metals.

13. The method of extracting tin from its ores was probably very defective in ancient times. At least, it was so for several centuries before the time of Elizabeth, when Sir Francis Godolphin introduced great improvements in the tin works. The use of the reverberatory furnace was commenced, about the beginning of the eighteenth century, and soon after pit-coal was substituted for charcoal.

14. This metal, in its solid state, is called *block-tin*. It is applied, without any admixture with any other metal, to the formation of vessels, which are not to be exposed to a temperature much above that of hot water. A kind of ware, called *biddery ware*, is made of tin alloyed with a little copper. The vessels made of this composition, are rendered black by the application of nitre, common salt, and sal ammoniac. *Foil* is also made by pressing it between steel rollers,

or by hammering it, as in the case of gold by the gold-beaters.

15. But tin is most extensively applied as a coating to other metals, stronger than itself, and more subject to oxydation. The plates which are usually denominated tin, are thin sheets of iron coated with this metal. The iron is reduced to thin plates in a rolling-mill, and these are prepared for being tinned, by first steeping them in water acidulated with muriatic acid, and then freeing them from oxyde by heating, scaling, and rolling them.

16. The tin is melted in deep oblong vessels, and kept in a state of fusion by a charcoal fire. To preserve its surface from oxydation, a quantity of fat or oil is kept floating upon it. The plates are dipped perpendicularly into the tin, and held there for some time. When withdrawn, they are found to have acquired a bright coating of the melted metal. The dipping is performed three times for *single tin plate*, and six times for *double tin plate*. The tin penetrates the iron, and forms an alloy.

17. Various articles of iron, such as spoons, nails, bridle-bits, and small chains, are coated with tin, by immersing them in that metal, while in a state of fusion. The great affinity of tin and copper, renders it practicable to apply a thin layer of the former metal to the surface of the latter; and this is often done, as stated in the article on the coppersmith.

18. Tin and quicksilver are applied to the polished surface of glass, for the purpose of forming mirrors. In silvering plain looking-glasses, a flat, horizontal slab is used as a table. This is first covered with paper, and then with a sheet of tin foil of the size of the glass. A quantity of quicksilver is next laid on the foil, and spread over it with a roll of cloth, or with a hare's foot.

19. After as much quicksilver as the surface will

hold, has been spread on, and while it is yet in a fluid state, the glass is shoved on the sheet of foil from the edge of the table, driving a part of the liquid metal before it. The glass is then placed in an inclined position, that every unnecessary portion of the quicksilver may be drained off, after which it is again laid flat upon the slab, and pressed for a considerable time with heavy weights. The remaining quicksilver amalgamates with the tin, and forms a permanent, reflecting surface.

THE TIN-PLATE WORKER.

1. THE materials on which the tinner, or tin-plate worker, operates, are the rolled sheets of iron, coated with tin, as just described. He procures the sheets by the box, and applies them to the roofs and other parts of houses, or works them up into various utensils, such as pails, pans, bake-ovens, measures, cups, and ducts for conveying water from the roofs of houses.

2. In making the different articles, the sheets are cut into pieces of proper size, with a huge pair of shears, and these are brought to the proposed form by different tools, adapted to the purpose. The several parts are united by means of a solder made of a composition of tin and lead. The solder is melted, and made to run to any part, at the will of the workman, by means of a copper instrument, heated for the purpose in a small furnace with a charcoal fire.

3. On examining almost any vessel of tin ware, it will be perceived, that, where the parts are united, one of the edges, at least, and sometimes both, are turned, that the solder may be easily and advantageously applied. It will also be discovered that iron wire is applied to those parts requiring more strength than is possessed by the tin itself. The edges and handles are especially strengthened in this manner.

4. The edges of the tin were formerly turned on a

steel edge, or a kind of anvil called a *stock*, with a mallet; and, in some cases, this method is still pursued; but this part of the work is now more expeditiously performed, by means of several machines invented by Seth Peck, of Hartford Co., Connecticut These machines greatly expedite the manufacture of tin wares, and have contributed much towards reducing their price.

5. This manufacture is an extensive branch of our domestic industry; and vast quantities of tin, in the shape of various utensils, are sold in different parts of the United States, by a class of itinerant merchants, called *tin-pedlers*, who receive in payment for their goods, rags, old pewter, brass, and copper, together with feathers, hogs' bristles, and sometimes ready money.

LEAD.

1. NEXT to iron, lead is the most extensively diffused, and the most abundant metal. It is found in various combinations in nature; but that mineralized by sulphur is the most abundant. This ore is denominated *galena* by the mineralogists, and is the kind from which nearly all the lead of commerce is extracted.

2. The ore having been powdered, and freed, as far as possible, from stony matter, is fused either in a blast or reverberatory furnace. In the smelting, lime is used as a flux, and this combines with the sulphur and earthy matters, while the lead unites with the carbon of the fuel, and sinks to the bottom of the furnace, whence it is occasionally let out into a reservoir.

3. Lead extracted from galena, often contains a sufficient proportion of silver to render it an object to extract it. This is done by oxydizing the lead by means of heat, and a current of air. At the end of this operation, the silver remains with a small quan

tity of lead, which is afterwards separated by the process of cupellation. The oxyde is applied to the purposes for which it is used, or it is reduced again to a metallic state.

4. The lead mines on the Mississippi are very productive, and very extensive. The principal mines are in the neighborhood of Galena, in the north-western part of Illinois, and these are the richest on the globe. The lead mines in the vicinity of Potosi, Missouri, are also very productive. About 3,000,000 pounds are annually smelted in the United States.

5. Lead, on account of its easy fusibility and softness, can be readily applied to a variety of purposes. It is cast in moulds, to form weights, bullets, and other small articles. Cisterns are lined, and roofs, &c.. are covered with sheet lead; and also in the construction of pumps and aqueducts, leaden pipes are considerably used. The mechanic who applies this metal to these purposes, is called a plumber.

6. Lead is cast into sheets in sand, on large tables having a high ledge on each side. The melted lead is poured out upon the surface from a box, which is made to move on rollers across the table, and is equalized, by passing over it a straight piece of wood called a *strike*. The sheets thus formed, are afterwards reduced in thickness, and spread to greater dimensions, by compressing them between steel rollers.

7. Leaden pipes may be made in various ways. They were at first formed of sheet lead, bent round a cylindrical bar, or mandrel, and then soldered; but pipes formed in this manner, were liable to crack and break. The second method consists in casting successive portions of the tube in a cylindrical mould, having in it a core. As soon as the tube gets cold, it is drawn nearly out of the mould, and more lead is poured in, which unites with the tube previously form-

ed. But pipes cast in this way are found to have imperfections, arising from flaws and air bubbles.

8. In the third method, which is the one most commonly practised, a thick tube of lead is cast upon one end of a long polished iron cylinder, or mandrel, of the size of the bore of the intended pipe. The lead is then reduced, and drawn out in length, either by drawing it on the mandrel through circular holes of different sizes, in a steel plate, or by rolling it between contiguous rollers, which have a semicircular groove cut round the circumference of each.

9. The fourth method consists in forcing melted lead, by means of a pump, into one end of a mould, while it is discharged in the form of a pipe, at the opposite end. Care is taken so to regulate the temperature, that the lead is chilled just before it leaves the mould.

10. *Shot* is likewise made of lead. These instruments of death are usually cast in high towers constructed for the purpose. The lead is previously alloyed with a small portion of arsenic, to increase the cohesion of its particles, and to cause it to assume more readily the globular form. It is melted at the top of the tower, and poured into a vessel perforated at the bottom with a great number of holes.

11. The lead, after running through these perforations, immediately separates into drops, which cool in falling through the height of the tower. They are received below in a reservoir of water, which breaks the fall. The shot are then proved by rolling them down a board placed in an inclined position. Those which are irregular in shape roll off at the sides, or stop, while the spherical ones continue on to the end.

II.—T

THE IRON-FOUNDER, &c.

IRON.

1. THE properties which iron possesses in its various forms, render it the most useful of all the metals. The toughness of *malleable iron* renders it applicable to purposes, where great strength is required, while its difficult fusibility, and property of softening by heat, so as to admit of forging and welding, cause it to be easily wrought.

2. Cast iron, from its cheapness, and from the facility with which its form may be changed, is made the material of numerous structures. *Steel*, which is the most important compound of iron, exceeds all other metals in hardness and tenacity; and hence it is particularly adapted to the fabrication of cutting instruments.

3. Iron was discovered, and applied to the purposes of the arts, at a very early period. Tubal-Cain, who was the seventh generation from Adam, "was an instructer of every artificer in brass and iron." Noah must have used much of this metal in the construction of the ark, and, of course, he must have transmitted a knowledge of it to his posterity.

4. Nevertheless, the mode of separating it from the various substances with which it is usually combined, was but imperfectly understood by the ancients; and their use of it was, most likely, confined chiefly to the limited quantity found in a state nearly pure. Gold, silver, copper, and tin, are more easily reduced to a state in which they are available in the arts. They were, therefore, often used in ancient times, for purposes to which iron would have been more applicable. This was the case especially with copper and tin.

5. Fifteen distinct kinds of iron ore, have been discovered by mineralogists; but of these, not more than four have been employed in making iron. There are, however, several varieties of the latter kind, all of which are classed by the smelters of iron under the general denomination of *bog* and *mountain* or *hard* ores.

6. The former has much of the appearance of red, brown, or yellowish earth, and is found in beds from one to six feet thick, and in size from one fourth of a rood to twenty acres. The mountain, or hard ore, to a superficial observer, differs but little in its appearance from common rocks or stones. It is found in regular strata in hills and mountains, or in detached masses of various sizes, and in hilly land from two to eight feet below the surface.

7. The bog-ore is supposed to be a deposite from water which has passed over the hard ore. This is evidently the case in hilly countries, where both kinds

occur. Some *iron-masters* use the bog; some, the hard; and others, both kinds together. In this particular, they are governed by the ore, or ores, which may exist in their vicinity.

8. The apparatus in which the ore is smelted, is called a *blast-furnace*, which is a large pyramidal stack, built of hewn stone or brick, from twenty to sixty feet in height, with a cavity of a proportionate size. In shape, this cavity is near that of an egg, with the largest end at the bottom. It is lined with fire-brick or stone, capable of resisting an intense heat.

9. Below this cavity is placed the *hearth*, which is composed of four or five large coarse sandstones, split out of a solid rock, and chiselled so as to suit each other exactly. These form a cavity for the reception of the iron and dross, when melted above. The hearth requires to be removed at the end of every *blast*, which is usually continued from six to ten months in succession, unless accidentally interrupted.

10. The preparation for a blast, consists principally in providing charcoal and ore. The wood for the former is cut in the winter and spring, and charred and brought to the furnace during the spring, summer, and autumn. What is not used during the time of hauling, is stocked in coal-houses, provided for the purpose.

11. The wood is charred in the following manner. It is first piled in heaps of a spherical form, and covered with leaves and dirt. The fire is applied to the wood, at the top, and when it has been sufficiently ignited, the pit is covered in; but, to support combustion, several air-holes are left near the ground. The *colliers* are obliged to watch the pit night and day, lest, by the caving in of the dirt, too much air be admitted, and the wood be thereby consumed to ashes.

12. When the wood has been reduced to charcoal, the fire is partially extinguished by closing the air-holes. The coals are *drawn* from the pit with an iron-toothed rake, and, while this is performed, the dust mingles with them, and smothers the fire which may yet remain. Wood is also charred in kilns made of brick.

13. The hard ore is dug by *miners*, or, as they are commonly denominated, *ore-diggers*. In the prosecution of their labor, they sometimes follow a vein into a hill or mountain. When the ore is found in strata or lumps near the surface, they dig down to it. This kind of ore commonly contains sulphur and arsenic, and to free it from those substances, and to render it less compact, it is roasted in kilns, with refuse charcoal, which is too fine to be used for any other purpose. It is then broken to a suitable fineness with a hammer, or in a crushing mill. The bog-ore seldom needs any reduction.

14. Every preparation having been made, the furnace is gradually heated with charcoal, and by degrees filled to the top, when a small quantity of the ore is thrown on, and the blast is applied at the bottom near the hearth. The blast is supplied by means of one or two cylindrical bellows, the piston of which is moved by steam or water power.

15. The coal is measured in baskets, holding about one bushel and a half, and the ore, in boxes holding about one peck. Six baskets of coal, and as many boxes of ore as the furnace can carry, is called a *half charge*, which is renewed as it may be necessary to keep the furnace full. With every charge is also thrown in one box of limestone.

16. The limestone is used as a flux, to aid in the fusion of the ore, and to separate its earthy portions from the iron. The iron sinks by its specific gravity, to the bottom of the hearth, and the earthy portions,

now converted into glass by the action of the limestone and heat, also sink, and float upon the liquid iron This scum, or, as it is usually called, scoria, slag, or cinder, is occasionally removed by instruments made for the purpose.

17. When the hearth has become full of iron, the metal is let out, at one corner of it, into a bed of sand, called a *pig-bed*, which is from twenty to thirty feet in length, and five or six in width. One concave channel, called *the sow*, extends the whole length of the bed, from which forty or fifty smaller ones, called *pig-moulds*, extend at right angles. The metal, when cast in these moulds, is called *pig-iron*, and the masses of iron, *pigs*.

18. *Pig-iron*, or, as it is sometimes called, *crude iron*, being saturated with carbon and oxygen, and containing also a portion of scoria, is too brittle for any other purpose than castings. Many of these, such as stoves, grates, mill-irons, plough-irons, and kitchen utensils, are commonly manufactured at blast furnaces, and in many cases nearly all the iron is used for these purposes. In such cases, the metal is taken in a liquid state, from the hearth, in ladles.

19. In Great Britain and Ireland, and perhaps in some other parts of Europe, iron-ore is smelted with *coke*, a fuel which bears the same relation to pit-coal, that charcoal does to wood. It is obtained by heating or baking the coal in a sort of oven or kiln, by which it becomes charred. During the process, a sort of bituminous tar is disengaged from it, which is carefully preserved, and applied to many useful purposes.

THE IRON-FOUNDER.

1. THE appellation of *founder* is given to the superintendent of a blast-furnace, and likewise to those persons who make castings either of iron or any other metal. In every case, the term is qualified by

a word prefixed, indicating the metal in which he operates, or the kind of castings which he may make; as *brass*-founder, *iron*-founder, or *bell*-founder. But what soever may be the material in which he operates, or the kind of castings which he may produce, his work is performed on the same general principle.

2. The sand most generally employed by the founder is *loam*, which possesses a sufficient proportion of argillaceous matter, to render it moderately cohesive, when damp. The moulds are formed by burying in the sand, wooden or metallic patterns, having the exact shape of the respective articles to be cast. To exemplify the general manner of forming moulds, we will explain the process of forming one for the *spider*, a very common kitchen utensil.

3. The pattern is laid upon a plain board, which in this application is called a *follow board*, and surrounded with a frame called a *flask*, three or four inches deep. This is filled with sand, and consolidated with rammers, and by treading it with the feet. Three wooden patterns for the legs are next buried in the sand, and a hole is made for pouring in the metal.

4. One side of the mould having been thus formed, the flask, with its contents, is turned over, and, the follow board having been removed, another flask is applied to the first, and filled with sand in the same manner. The two flasks are then taken apart, and the main pattern, together with those for the legs, removed. The whole operation is finished by again closing the flasks.

5. The mode of proceeding in forming moulds for different articles, is varied, of course, to suit their conformation. The pattern is often composed of several pieces, and the number and form of the flasks are also varied accordingly. Cannon-balls are sometimes cast in moulds of iron; and to prevent the melted metal from adhering to them, the inside is covered with pulverized black lead

6. Rollers for flattening iron are also cast in iron moulds. This method is called *chill-casting*, and has for its object the hardening of the surface of the metal, by the sudden reduction of the temperature, which takes place in consequence of the great power of the mould, as a conductor of heat. These rollers are afterwards turned in a powerful lathe.

7. Several *moulders* work together in one foundery, and, when they have completed a sufficient number of moulds, they fill them with the liquid metal. The metal for small articles is dipped from the hearth or crucible of the furnace with iron ladles defended on every side with a thin coating of clay mortar, and poured thence into the moulds. But in casting articles requiring a great amount of iron, such as cannon, and some parts of the machinery for steam engines, the iron is transferred to the moulds, in a continued stream, through a channel leading from the bottom of the crucible. In such cases, the moulds are constructed in a pit dug in the earth near the furnace. Large ladles full of iron are, in some founderies, emptied into the moulds by the aid of huge cranes.

8. Although the moulders have their distinct work to perform, yet they often assist each other in lifting heavy flasks, and in all cases, in filling the moulds. The latter operation is very laborious; but the exertion is continued but a short time, since the moulds, constructed during a whole day, can be filled in ten or fifteen minutes.

9. Iron-founderies are usually located in or near large cities or towns, and are supplied with crude iron, or pig metal, from the blast furnaces in the interior. The metal is fused either with charcoal or with pit coal. In the former case, an artificial blast is necessary to ignite the fuel; but in the latter, this object is often effected in air furnaces, which are so constructed that a sufficient current of air is obtained directly from the atmos here.

10. The practice of making castings of iron is comparatively modern; those of the ancients were made of brass, and other alloys of copper. Until the beginning of the last century, iron was but little applied in this way. This use of it, however, has extended so rapidly, that cast iron is now the material of almost every kind of machinery, as well as that of innumerable implements of common application. Even bridges and rail-roads have been constructed of cast iron.

THE BAR IRON MAKER.

1. BAR-IRON is manufactured from pig-iron, from *blooms*, and directly from the ore; the process is consequently varied in conformity with the state of the material on which it is commenced.

2. In producing bar-iron from pigs, the latter are melted in a furnace similar to a smith's forge, with a sloping cavity ten or twelve inches below, where the blast-pipe is admitted. This hearth is filled with charcoal and dross, or scoria; and upon these is laid the metal and more coal. After the coal has become well ignited, the blast is applied. The iron soon begins to melt, and as it liquefies, it runs into the cavity or hearth below. Here, being out of the reach of the blast, it soon becomes solid.

3. It is then taken out, and fused again in the same manner, and afterwards a third time. After the third heat, when the iron has become solid enough to bear beating, it is slightly hammered with a sledge, to free it from the adhering scoria. It is then returned to the furnace; but, being placed out of the reach of the blast, it soon becomes sufficiently compact to bear the *tilt-hammer*.

4. With this instrument, the iron is beaten, until the mass has been considerably extended, when it is cut into several pieces, which, by repeated beating and forging, are extended into bars, as we see them

for sale. The tilt-hammer weighs from six to twelve hundred pounds, and is most commonly moved by water power.

5. For manufacturing bar-iron directly from the ore, the furnace is similar in its construction to the one just described, and the operations throughout are very similar. A fire is first made upon the hearth with charcoal; and, when the fuel has become well ignited, a quantity of ore is thrown upon it, and the ore and the fuel are renewed as occasion may require. As the iron melts, and separates from the earthy portions of the ore, it sinks to the bottom of the hearth. The scoria is let off occasionally, through holes made for the purpose. When iron enough has accumulated to make a *loop*, as the mass is called, it is taken out, and forged into bars under the tilt-hammer.

6. This way of making bar-iron is denominated the *method of the Catalan forge*, and is by far the cheapest and most expeditious. It is in general use in all the southern countries of Europe, and it is beginning to be extensively practised in the United States. When a Catalan forge is employed in making *blooms*, it is called a *bloomery*.

7. The blooms are about eighteen inches long, and four in diameter. They are formed under the tilt-hammer, and differ in substance from bar-iron in nothing, except that, having been imperfectly forged, the fibres of the metal are not fully extended, nor firmly united. The blooms are manufactured in the interior of the country, where wood is abundant, and sold by the ton, frequently, in the cities, to be converted into bar or sheet iron.

8. These blooms are converted into bar-iron, by first heating them in an air-furnace, by means of stone coal, and then passing them between chill cast iron rollers. The rollers are filled with grooves, which gradually decrease in size from one side to the other.

When the iron has passed through these, the bloom of eighteen inches in length, has become extended to nearly as many feet. The bar thus formed, having been cut into four pieces, the process is finished by welding them together laterally, and again passing them between another set of rollers, by which they are brought to the form in which they are to remain.

9. Blooms are also laminated into two sheets, on the same principle, between smooth rollers, which are screwed nearer to each other every time the bloom is passed between them. Very thin plates, like those which are tinned for the tin-plate workers, are repeatedly doubled, and passed between the rollers, so that in the thinnest plates, sixteen thicknesses are rolled together, oil being interposed to prevent their cohesion. The last rollings are performed while the metal is cold.

10. Rolled plates of iron are frequently cut into rods and narrow strips. This operation is performed by means of elevated angular rings upon rollers, which are so situated that they act reciprocally upon each other, and cut like shears. These rings are separately made, so that they can be removed for the purpose of sharpening them, when necessary. The mills in which the operations of rolling and slitting iron are performed, are called rolling and slitting mills.

THE WIRE DRAWER.

1. IRON is reduced to the form of wire by drawing rods of it through conical holes in a steel plate. To prepare the metal for the operation of drawing, it is subjected to the action of the hammer, or to that of rollers, until it has been reduced to a rod sufficiently small to be forced through the largest hole. The best wire is produced from rods formed by the method first mentioned.

2. Various machines are employed to overcome the resistance of the plate to the passage of the wire. In general, the wire is held by pinchers, near the end, and as fast as it is drawn through the plate, it is wound upon a roller, by the action of a wheel and axle, or other power. Sometimes, a rack and pinion are employed for this purpose, and sometimes a lever, which acts at intervals, and which takes fresh hold of the wire every time the force is applied.

3. The finer kinds of wire are made from the larger by repeated drawings, each of which is performed through a smaller hole than the preceding. As the metal becomes stiff and hard, by the repetition of this process, it is occasionally annealed, to restore its ductility. Wire is formed of other metals by the same general method.

THE STEEL MANUFACTURER.

1. STEEL is a compound of iron and carbon; and, as there are several methods by which the combination is produced, there are likewise several kinds of steel. The best steel is said to be made of Swedish or Russian bar-iron.

2. The most common method of forming steel is by the process of *cementation.* The operation is performed in a conical furnace, in which are two large cases or troughs, made of fire-brick, or good fire stone; and beneath these is a long grate. On the bottom of the cases is placed a layer of charcoal dust, and over this a layer of bar-iron. Alternate strata of these materials are continued to a considerable height, ten or twelve tons of iron being put in at once.

3. The whole is covered with clay or sand, to exclude the air, and flues are carried through the pile from the furnace below, so as to heat the contents equally and completely. The fire is kindled in the

grate, and continued for eight or ten days, during which time, the troughs, with their contents, are kept red hot. The progress of the cementation is discovered by drawing a *test* bar from an aperture in the side.

4. When the conversion of the iron into steel appears to be complete, the fire is extinguished; and, after having been suffered to cool for six or eight days, it is removed. Iron combined with charcoal in this manner, is denominated *blistered steel*, from the blisters which appear on its surface, and in this state, it is much used for common purposes.

5. To render this kind of steel more perfect, the bars are heated to redness, and then drawn out into bars of much smaller dimensions, by means of a hammer moved by water or steam power. This instrument is called a tilting hammer, and the bars formed by it, are called *tilted steel*. When the bars have been exposed to heat, and afterwards doubled, drawn out, and welded, the product is called *shear steel*.

6. But steel of cementation, however carefully made, is never quite equable in its texture. Steel possessing this latter quality is made, by fusing bars of blistered steel, in a crucible placed in a wind furnace. When the fusion has been completed, the liquid metal is cast into small bars or ingots, which are known in commerce by the name of *cast steel*. Cast steel is harder, more elastic, closer in texture, and capable of receiving a higher polish than common steel.

7. Steel is also made directly from cast iron, or at once from the ore. This kind is called *natural* or *German* steel, and is much inferior to that obtained by cementation. The best steel, produced directly from the ore, comes from Germany, and is made in Stiria. It is usually imported in barrels, or in chests about three feet long.

8. Steel is sometimes alloyed with other metals. A celebrated Indian steel, called *wootz*, is supposed to be carbonated iron, combined with small quantities of silicium and aluminum. Steel alloyed with a very small proportion of silver, is superior to wootz, or to the best cast steel. Some other metals are also used with advantage in the same application.

9. Steel was discovered at a very early period of the world, for aught we know, long before the flood. Pliny informs us, that, in his time, the best steel came from China, and that the next best came from Parthia. A manufactory of steel existed in Sweden as early as 1340 of the Christian era: but it is generally thought, that the process of converting iron into steel by cementation originated in England, at a later period. The method of making cast steel was invented at Sheffield, in the latter country, in 1750, and, for a long time, it was kept secret.

10. It has been but a few years, since this manufacture was commenced in the United Sates. In 1836, we had fourteen steel furnaces, viz.; at Boston, one; New-York, three; Troy, one; New-Jersey, two; Philadelphia, three; York Co., Pa., one; Baltimore, one; and Pittsburg, two. These furnaces together are said to be capable of yielding more than 1600 tons of steel in a year. The American steel is employed in the fabrication of agricultural utensils, and it has entirely excluded the common English blistered steel.

THE BLACKSMITH, AND THE NAILER.

THE BLACKSMITH.

1. The blacksmith operates in wrought iron and steel, and, from these materials, he fabricates a great variety of articles, essential to domestic convenience, and to the arts generally.

2. This business is one of those trades essential in the rudest state of society. Even the American Indians are so sensible of its importance, that they cause to be inserted in the treaties which they make with the United States, an article stipulating for a blacksmith to be settled among them, and for a supply of iron.

3. The utility of this trade will be further manifest by the consideration, that almost every other business is carried on by its aid. The agriculturist is dependent on it for farming utensils, and mechanics and art-

ists of every description, for the tools with which they operate; in short, we can scarcely fix upon a single utensil, vehicle, or instrument, which does not owe its origin, either directly or indirectly, to the blacksmith.

4. This business being thus extensive in its application, it cannot be presumed that any one person can be capable of executing every species of work. This, however, is not necessary, since the demand for particular articles is frequently so great, that the whole attention may be directed to the multiplication of individuals of the same kind. Some smiths make only anchors, axes, scythes, hoes, or shovels.

5. In such cases, the workmen acquire great skill and expedition in the manufacture. A tilt hammer is often used in forging large masses of iron, and even in making utensils as small as the hoe, the axe, and the sword; but the hammer which may be employed bears a due proportion in its weight to the mass of iron to be wrought. In all cases in which a tilt hammer is used, the bellows from which the blast proceeds is moved by water or steam power.

6. In the shop represented at the head of this article, sledges and hammers are used as forging instruments, and these are wielded by the workmen themselves. The head workman has hold of a piece of iron with a pair of tongs, and he, with a hammer, and two others, with each a sledge, are forging it upon an anvil. The two men are guided in their disposition of the strokes chiefly by the hammer of the master-workman.

7. In ordinary blacksmith shops, two persons commonly work at one forge, one of whom takes the lead in the operations, and the other works the bellows, and uses the sledge. From the part which the latter takes in the labor, he is called the *blower* and *striker*. A man or youth, who understands but little of the business, can, in many cases, act in this capacity tolerably well.

8. The iron is rendered malleable by heating it with charcoal or with stone coal, which is ignited intensely by means of a blast from a bellows. The iron is heated more or less, according to the particular object of the workman. When he wishes to reduce it into form, he raises it to a *white heat*. The *welding heat* is less intense, and is used when two pieces are to be united by *welding*. At a red heat, and at lower temperatures, the iron is rendered more compact in its internal texture, and more smooth upon its surface.

9. The joint action of the heat and air, while the temperature is rising, tends to produce a rapid oxydation of the surface. This result is measurably prevented by immersing the iron in sand and common salt, which, uniting, from a vitreous coating for its protection. This coating is no inconvenience in the forging, as its fluidity causes it to escape immediately under the action of the hammer.

10. Steel is combined with iron in the manufacture of cutting instruments, and other implements, as well as articles requiring, at certain parts, a great degree of hardness. This substance possesses the remarkable property of changing its degree of hardness by the influence of certain degrees of temperature. No other substance is known to possess this property; but it is the peculiar treatment which it receives from the workman that renders it available.

11. If steel is heated to redness, and suddenly plunged into cold water, it is rendered extremely hard, but, at the same time, too brittle for use. On the other hand, if it is suffered to cool gradually, it becomes too soft and ductile. The great object of the operator is to give to the steel a quality equally distant from brittleness and ductility. The treatment by which this is effected is called *tempering*, which will be more particularly treated in the article on the cutler, whose employment is a refined branch of this business.

THE NAILER.

1. NAIL-MAKING constitutes an extensive branch of the iron business, as vast quantities of nails are annually reuqired by all civilized communities. They are divided into two classes, the names of which indicate the particular manner in which they are manufactured; viz., *wrought nails* and *cut nails*.

2. The former are usually forged on the anvil, and when a finished head is required, as is commonly the case, it is hammered on the larger end, after it has been inserted into a hole of an instrument formed for the purpose. Workmen by practice acquire surprising dispatch in this business; and this circumstance has prevented the general introduction of the machines which have been invented for making nails of this description. Wrought nails can be easily distinguished from cut nails, by the indentations of the hammer which have been left upon them.

3. In making cut nails, the iron is first brought into bars between grooved rollers. The size of the bars is varied in conformity with that of the proposed nails. These bars are again heated, and passed between smooth rollers, which soon spread them into thin strips of suitable width and thickness. These strips, having been cut into pieces two or three feet in length, are heated to a red heat in a furnace, to be immediately converted into nails, when designed for those of a large size. For small nails, the iron does not require heating.

4. The end of the plate is presented to the machine by the workman, who turns the material over, first one way and then the other; and at each turn a nail is produced. The machine has a rapid reciprocating motion, and cuts off, at every stroke, a wedge-like piece of iron, constituting a nail without a head. This is immediately caught near the head, and com-

pressed between *gripes;* and, at the same time, a force is applied to a die at the end, which spreads the iron sufficiently to form the head. From one to two hundred can be thus formed in a minute. This fact accounts for the low rate at which cut nails are now sold, which, on an average, is not more than two cents per pound above that of bar iron.

5. On account of the greater expense of manufacturing wrought nails, they are sold much higher. It is said that nine-tenths of all the nails of this kind, used in the United States, are imported from Europe. We thus depend upon foreign countries for these and many other articles, because they can be imported cheaper than we can make them; and this circumstance arises chiefly from the difference in the price of labor.

6. The first machine for making cut nails was invented in Massachusetts about the year 1816, by a Mr. Odion, and soon afterwards another was contrived, by a Mr. Reed, of the same state. Other machines, for the same purpose, have likewise been constructed by different persons, but those by Odion and Reed are most commonly used. Before these machines were introduced, the strips of iron just described, were cut into wedgelike pieces by an instrument which acted on the principle of the shears; and these were afterwards headed, one by one, with a hammer in a vice. The fact, that the manufacture of this kind of nails originated in our country, is worthy of recollection.

7. In 1841, Walter Hunt, of New-York, invented a double reciprocating nail engine, which is owned by the New-York Patent Nail Company. This machine works with surprising rapidity, it being capable of cutting five or six hundred ten-penny nails in a minute. One hand can tend three engines, as he has nothing more to do than to place the heated plate in a perpendicular position in the machine.

8. This manufacture includes, also, that of tacks and spikes; but since, in the production of these, the same general methods are pursued, they need no particular notice. The different sizes of tacks are distinguished by a method which indicates the number per ounce; as two, three, or four hundred per ounce. Spikes are designated by their length in inches, and nails by the terms, two-penny, three-penny, four-penny, ten-penny, and so on up to sixty-penny.

THE CUTLER.

1. UNDER the head of cutlery, is comprehended a great variety of instruments designed for cutting and penetration, and the business of fabricating them is divided into a great number of branches. Some manufacture nothing but axes; others make plane-irons and chisels, augers, saws, or carvers' tools. Others, again, make smaller instruments, such as table-knives, forks, pen-knives, scissors, and razors. There are also cutlers who manufacture nothing but surgical instruments.

2. The coarser kinds of cutlery are made of blistered steel welded to iron. Tools of a better quality are made of shear steel, while the sharpest and most delicate instruments are formed of cast steel. The several processes constituting this business may be comprised in forging, tempering, and polishing; and

these are performed in the order in which they are here mentioned.

3. The general method of *forging* iron and steel, in every branch of this business, is the same with that used in the common blacksmith's shop, for more ordinary purposes. The process, however, is somewhat varied, to suit the particular form of the object to be fashioned; for example, the blades and some other parts of the scissors are formed by hammering the steel upon indented surfaces called *bosses*. The bows, which receive the finger and thumb, are made by first punching a hole in the metal, and then enlarging it by the aid of a tool called a *beak-iron*.

4. The steel, after having been forged, is soft, like iron, and to give it the requisite degree of strength under the uses to which the tools or instruments are to be exposed, it is hardened. The process by which this is effected is called *tempering*, and the degree of hardness or strength to which the steel is brought is called its *temper*, which is required to be *higher* or *lower*, according to the use which is to be made of the particular instrument.

5. In giving to the different kinds of instruments the requisite temper, they are first heated to redness, and then plunged into cold water. This, however, raises the temper too high, and, if left in this condition, they would be too brittle for use. To bring them to a proper state, they are heated to a less degree of temperature, and again plunged into cold water. The degree to which they are heated, the second time, is varied according to the hardness required. That this particular point may be perfectly understood, a few examples will be given.

6. Lancets are raised to 430 degrees Fahrenheit. The temperature is indicated by a pale color, slightly inclined to yellow. At 450 degrees, a pale straw-color appears, which is found suitable for the best razors and surgical instruments. At 470 degrees, a full ---

or is produced, which is suitable for pen-knives, common razors, &c. At 490, a brown color appears, which is the indication of a temper proper for shears, scissors, garden hoes, and chisels intended for cutting cold iron.

7. At 510 degrees, the brown becomes dappled with purple spots, which shows the proper heat for tempering axes, common chisels, plane-irons, &c. At 530 degrees, a purple color is established, and this temperature is proper for table-knives and large shears. At 550 degrees, a bright blue appears, which is proper for swords and watch springs. At 560 degrees, the color is full blue, and this is used for fine saws, augers, &c. At 600 degrees, a dark blue approaching to black settles upon the metal, and this produces the softest of all the grades of temper, which is used only for the larger kinds of saws.

8. Other methods of determining the degree of temperature at which the different kinds of cutlery are to be immersed, a second time, in cold water, are also practised. By one method, the pieces of steel are covered with tallow or oil, or put into a vessel containing one of these substances, and heated over a moderate fire. The appearance of the smoke indicates the degree of heat to which it may have been raised. A more accurate method is found in the employment of a fluid medium, the temperature of which can be regulated by a thermometer. Thus oil, which boils at 600 degrees, may be employed for this purpose, at any degree of heat which is below that number.

9. The *grinding* of cutlery is effected on cylindrical stones of various kinds, among which freestone is the most common. These are made to revolve with prodigious velocity, by means of machinery. The operation is therefore quickly performed. The *polishing* is commonly effected by using, first, a wheel of wood; then, one of pewter; and, lastly, one covered with buff leather sprinkled with an impure oxyde of

iron, called *colcothar* or *crocus*. The edges are set either with hones or whetstones, or with both, according to the degree of keenness required.

10. Almost every description of cutlery requires a handle of some sort; but the nature of the materials, as well as the form and mode of application, will be readily understood by a little attention to the various articles of this kind which daily fall in our way.

11. A process has been invented, by which edge tools, nails, &c., made of cast iron, may be converted into good steel. It consists in stratifying the articles with the oxyde of iron, in a metallic cylinder, and then submitting the whole to a regular heat, in a furnace built for the purpose. This kind of cutlery, however, will not bear a very fine edge.

12. The sword and the knife were probably the first instruments fabricated from iron, and they still continue to be leading subjects of demand, in all parts of the world. The most celebrated swords of antiquity were made at Damascus, in Syria. These weapons never broke in the hardest conflicts, and were capable of cutting through steel armor without sustaining injury.

13. The fork, as applied in eating, is an invention comparatively modern. It appears to have had its origin in Italy, probably in the fourteenth century; but it was not introduced into England, until the reign of James the First, in the first quarter of the seventeenth. Its use was, at first, the subject of much ridicule and opposition.

14. Before the introduction of the fork, a piece of paper, or something in place of it, was commonly wrapped round some convenient projection of the piece to be carved; and, at this place, the operator placed one hand, while he used the knife with the other. The carver cut the mass of meat into slices or suitable portions, and laid them upon the large slices of bread which had been piled up near the platter, or

carving dish, and which, after having been thus served, were handed about the table, as we now distribute the plates.

15. The knives used at table were pointed, that the food might be taken upon them, as upon a fork; and knives of the same shape are still common on the continent of Europe. Round-topped knives were not adopted in Paris, until after the banishment of Napoleon Bonaparte to Elba, in 1815, when every thing English became fashionable in that city.

16. In France, before the revolution of 1789, it was customary for every gentleman, when invited to dinner, to send his knife and fork before him by a servant; or, if he had no servant, he carried them himself in his breeches pocket. A few of the ancient regime still continue the old custom. The peasantry of the Tyrol, and of some parts of Germany and Switzerland, generally carry about them a case, containing a knife and fork, and a spoon.

17. The use of the fork, for a long time, was considered so great a luxury, that the members of many of the monastic orders were forbidden to indulge in it. The Turks and Asiatics use no forks, even to this day. The Chinese employ, instead of this instrument, two small sticks, which they hold in the same hand, between different fingers.

18. The manufacture of cutlery is carried on most extensively in England, at Birmingham, Sheffield, Walsall, Wolverhampton, and London. London cutlery has the reputation of being the best, and this circumstance induces the dealers in that city, to affix the London mark to articles made at other places. In the United States, there are many establishments for the fabrication of the coarser kinds of cutlery, such as axes, plane-irons, saws, hoes, scythes, &c., but for the finer descriptions of cutting instruments, we are chiefly dependent on Europe.

THE GUN-SMITH.

1. It is the business of the gun-smith to manufacture fire-arms of the smaller sorts; such as muskets, fowling-pieces, rifles, and pistols.

2. The principal parts of the instruments fabricated by this artificer, are the barrel, the stock, and the lock. In performing the operations connected with this business, great attention is paid to the division of labor, especially in large establishments, such as those belonging to the United States, at Springfield and Harper's Ferry; for example, one set of workmen forge the barrels, ramrods, or some part of the lock; others reduce some part of the forged material to the exact form required, by means of files; and again another class of operators perform some part of the work relating to the stock.

3. The barrel is formed by forging a bar of iron into a flat piece of proper length and thickness, and by turning the plate round a cylindrical rod of tempered steel, called a *mandril*, the diameter of which is considerably less than the intended bore of the barrel. The edges of the plate are made to overlap each other about half an inch, and are welded together by heating the tube in lengths of two or three inches at a time, and by hammering them with very brisk, but moderate strokes, upon an anvil which has a number of semicircular furrows upon it.

4. In constructing barrels of better workmanship, the iron is forged in smaller pieces, eight or nine inches long, and welded together laterally, as well as lengthwise. The barrel is now finished in the usual way; or it is first made to undergo the additional operation of *twisting*, a process employed upon those intended to be of superior quality. The operation is performed by heating small portions of it at a time, and twisting them successively, while one end is held fast.

5. The barrel is next bored with several bits, each a little larger than the preceding one. The last bit is precisely the size of the intended calibre. After the barrel has been polished, and the breech closed with a screw, its strength and soundness are tested by means of a ball of the proper size, and a charge of powder equal in weight to the ball. Pistol-barrels, which are to go in pairs, are forged in one piece, which is cut asunder, after it has been bored.

6. Barrels for rifles are much thicker than those for other small arms; and, in addition to the boring in common barrels, they are furrowed with a number of grooves or *rifles*, which extend from one end of the cavity to the other, either in a straight or spiral direction. These rifles are supposed to prevent the rolling

of the ball in its passage out, and to direct it more unerringly to the object of aim.

7. The stocks are commonly manufactured from the wood of the walnut-tree. These are first dressed in a rough manner, usually in the country. After the wood has been properly seasoned, they are finished by workmen, who commonly confine their attention to this particular branch of the business. In each of the United States' armories, is employed a machine with which the stocks are turned, and also one, with which the place for the lock is made.

8. The several pieces composing the lock are forged on anvils, some of which have indented surfaces, the more readily to give the proposed form. They are reduced somewhat with the file, and polished with substances usually employed for such purposes. The several pieces of the lock having been put together, it is fastened to the stock with screws. Other particulars in regard to the manufacture of small-arms will be readily suggested by a careful inspection of the different kinds, which are frequently met with.

9. The period at which, and the country where, gunpowder and fire-arms were first invented, cannot be certainly determined. Some attribute their invention to the Chinese; and, in confirmation of this opinion, assert that there are now cannon in China, which were made in the eightieth year of the Christian era. On this supposition, their use was gradually extended to the West, until they were finally adopted in Europe, in the fourteenth century.

10. Others, however, attribute the invention of gunpowder to Berthold Schwartz, a monk, who lived at Mentz, between the years 1290 and 1320. It is said, that in some of his alchemistic experiments, he put some saltpetre, sulphur, and charcoal, into a mortar, and having accidentally dropped into it a spark of fire, the contents exploded, and threw the pestle into

the air. This circumstance suggested to his mind the employment of the mixture for throwing projectiles. Some traditions, however, attribute the invention to Constantine Antlitz, of Cologne.

11. The fire-arms first used in Europe were cannon, and these were originally made of wood, wrapped in numerous folds of linen, and well secured with iron hoops. They were conical in shape, being widest at the muzzle; but this form was soon changed for the cylindrical. At length they were made of bars of iron, firmly bound together with hoops of the same metal. In the second half of the fourteenth century, a composition of copper and tin, which was brought to form by casting in sand, came into use.

12. Cannon were formerly dignified with great names. Charles V. of Spain had twelve, which he called after the *twelve apostles*. One at Bois-le-Duc is called the *devil*; a sixty-pounder, at Dover Castle, is called *Queen Elizabeth's pocket-pistol*; an eighty-pounder, at Berlin, is called the *thunderer*; two sixty-pounders, at Bremen, the *messengers of bad news*. But cannon are, at present, denominated from the weight of the balls which they carry; as six-pounders, eight-pounders, &c.

13. Fire-arms of a portable size were invented, about the beginning of the sixteenth century. The musket was the first of this class of instruments that appeared, and the Spanish nation, the first that adopted its use as a military weapon. It was originally very heavy, and could not be well supported in a horizontal position without a *rest*. The soldiers, on their march, carried only the rest and ammunition, while each was followed by a youth who bore the musket.

14. The powder was not ignited with a spark from a flint, but with a match. Afterwards, a lighter match-lock musket was introduced, which was car-

ried by the soldiers themselves. The rest, however, maintained its ground, until about the middle of the seventeenth century. The troops throughout Europe were furnished with fire-locks, such as are now used, a little before the beginning of the eighteenth century.

15. The bayonet was invented, about the year 1340, at Bayonne; but it was not generally introduced, until the pike was entirely discontinued, about sixty years afterwards. It was first carried by the side, and was used as a dagger in close fight; but, in 1690, the custom of fastening it to the muzzle of the fire-lock was commenced in France, and the example was soon followed throughout Europe.

16. Gunpowder, on which the use of fire-arms depends, is a composition of salt-petre, sulphur, and charcoal. The proportion of the ingredients is varied considerably in different countries, and by different manufacturers in the same country. But good gunpowder may be made of seventy-six parts of saltpetre, fifteen of charcoal, and nine of sulphur. These materials are first reduced to a fine powder separately, and then formed into a homogeneous mass by moistening the mixture with water, and pounding it for a considerable time in wooden mortars.

17. After the paste has been suffered to dry a little, it is forced through a kind of sieve. By this process it is divided into grains, the size of which depends upon that of the holes through which they have been passed. The powder is then dried in ovens, and afterwards put into barrels, which are made to revolve on their axis. The friction produced by this motion destroys the asperities of the grains, and renders their surfaces smooth and capable of easy ignition.

THE VETERINARY SURGEON.

1. THE horse, as well as the other domestic animals, is subject to a great variety of diseases, which, like those affecting the human system, are frequently under the control of medicinal remedies; and the same general means which are efficacious in healing the disorders of our race, are equally so in controlling those of the inferior part of the animal creation.

2. The great value of the domestic animals has rendered them, from the earliest periods, the objects of study and attention, not only while in health, but also when laboring under disease. For the latter state, a peculiar system was early formed, including a *materia medica*, and a general mode of treatment considerably different from those for human patients.

3. Of the authors of this system, whether Greek or

Roman, nothing worthy of notice has been transmitted to us, beyond an occasional citation of names, in the works of Columella, a Roman writer, who flourished in the reign of Tiberius Cæsar, and in Vegetius Renatus, who lived two centuries afterwards. The former treated at large on the general management of domestic animals, and the latter more professedly on the diseases to which they are liable.

4. Both of these writers treated their subject in elegant classical Latin ; but neither they nor any other ancient author whose works have reached us, had any professional acquaintance with medicine or surgery. Celsus is the only physician of those times who is said to have written on animal medicine ; but this part of his works is not extant.

5. Xenophon is the oldest veterinary writer whose work remains ; but his treatise is confined to the training and management of the horse for war and the chase. The chief merit of the ancient writers on this subject consists in the dietetic rules and domestic management which they propose. Their medical prescriptions are said to be an inconsistent and often discordant jumble of many articles, devoid of rational aim or probable efficacy.

6. On the revival of learning in Europe, when the anatomy and physiology of the human body had become grand objects of research in the Italian schools, veterinary anatomy attracted the attention of Ruini and others, whose descriptive labors on the body of the horse have since served for the ground-work and model to all the schools in Europe.

7. The works of the veterinary writers of antiquity were eagerly sought and translated in Italy and France, and the art was extensively cultivated, sometimes under regular medical professors. Every branch of the equine economy was pursued with assiduity and success, whether it related to harness and trappings, equi-

tation and military menage, or the methodical treatment of the hoof, and the invention of various kinds of iron shoes. Evangelista of Milan distinguished himself in the education or breaking of the horse; and to him is attributed the invention of the martingale.

8. The new science having been extended over a great proportion of the continent of Europe, could scarcely fail of occasional communication with England; nevertheless, the medical treatment of horses and other domestic animals continued exclusively in the hands of farriers and cow-doctors, until some time in the first quarter of the eighteenth century.

9. At this period, that branch of this art which relates to the medical and surgical treatment of the horse, attracted the attention of William Gibson, who had acted in the capacity of army surgeon in the wars of Queen Anne. He was the first author of the regular medical profession, in England, who attempted to improve veterinary science; and the publication of his work forms an era in its annals, since his work became, and has continued to the present day, the basis of the superior practice of the English.

10. The eighteenth century was abundantly fruitful in veterinary pursuits and publications. France took the lead; but a zeal for this branch of science pervaded Germany and the states north of that part of Europe, and colleges were established in various countries, with the express view of cultivating this branch of the medical art. It is said that the French have improved the anatomical and surgical branches of the art, and the English, those which relate to the application of medicines.

11. The first veterinary school was instituted at Lyons, in 1762. Another was established at Alfort, in 1766. A similar institution was opened at Berlin, in 1792, and in the same year, one at St. Pancras, near London. In these colleges, lectures are given,

and degrees conferred. In the diplomas, the graduate is denominated *veterinary surgeon*. A great number of these surgeons have been dispersed in the armies of Europe, as well as through the different countries, where they have been employed in the medical and surgical treatment of diseased animals, to the great advantage of their owners.

12. From the preceding account, it is evident, that the light of science has shone conspicuously, in Europe, on the domestic animals, in relation to their treatment, both while in health, and when laboring under disease. In the United States, we have no institution for the cultivation of this branch of knowledge. The press, however, has been prolific in the production of works treating on the various branches of the veterinary art; and many persons, by their aid, have rendered themselves competent to administer to animals in cases of disease, in a rational manner.

13. Nevertheless, the practice of animal medicine is confined chiefly to illiterate men, who, from their laborious habits, or from other causes, have not attained to that degree of information on animal diseases, and the general effects of medicine, that might enable them to prescribe their remedies on scientific principles. But this state of things is not peculiar to our country; for, notwithstanding the laudable efforts of enlightened men in Europe, the blacksmiths form a vast majority of the horse-surgeons and physicians in every part of it; and the medical treatment of the other domestic animals is commonly intrusted to persons who are still more incompetent.

14. The attention of blacksmiths was very early turned to the diseases of the horse, from the practice of supplying him with shoes. The morbid affections of the foot were probably the first which attracted their notice; and descanting upon these induced the general

belief, that they understood every other disease which might affect the animal.

15. These men, as artificers in iron, were orginally termed ferrers or ferriers, from the Latin word *ferrum*, iron; and their craft, ferriery. These terms, by a usual corruption or improvement in language, have been changed to farrier and farriery, both of which still remain in general use, the former as applied to persons who shoe horses and administer to them medicines and surgical remedies, and the latter to the art itself, by which they are, or ought to be, guided.

16. The appellation of veterinary surgeon is applicable to persons who have received a diploma from some veterinary college, or who have, at least, studied animal medicine scientifically. There are a few such individuals in the United States; and the great value of the domestic animals, and the general increase of knowledge, certainly justify the expectation, that their number will increase.

THE END.

www.ingramcontent.com/pod-product-compliance
Lightning Source LLC
Chambersburg PA
CBHW031951230426
43672CB00010B/2120